Statistics for Business Decisions

MTH 251-252

Elaine A. Previte, MBA, MA with Dominic Nocera, MA, MS
Bristol Community College

Academx

Statistics for Business Decisions – MTH 251-252

Permissions Department
Academx Publishing Services, Inc.
P.O. Box 208
Sagamore Beach, MA 02562
http://www.academx.com

Printed in the United States of America

ISBN-13: 978-1-68284-307-9
ISBN-10: 1-68284-307-6

TABLE OF CONTENTS

TABLES

Introduction to Business Statistics
Chapter 1 Summary

Why Learn Statistics?

Statistics is an important area of contemporary mathematics – some would argue that is the most relevant subject area for the general population; but for business people it is a critical subject. Businesses are inundated with huge amounts of data, so there is a need to synthesize that data to put it in a usable form.

Most information is conveyed using the vocabulary of statistics (see below), so a statistically literate person must be conversant with this lingo. Additionally, businesses need to use information in all areas of decision-making, so knowing how to make proper inferences from data is essential to a successful enterprise.

Statistics in Business

Business people use the two key areas in statistics:

- *Descriptive statistics* are used when collecting, summarizing, presenting, and analyzing large data sets.
- *Inferential statistics* are used when estimating population values based on the statistics obtained from smaller data sets, and when analyzing relationships between different variables to make predictions concerning various behaviors.

Basic Vocabulary of Statistics

Several key terms must become part of your everyday vocabulary.

- *Variable:* a characteristic of an item or an individual
- *Data:* different values associated with a variable
- *Population:* all the items or individuals about which you wish to reach conclusions
- *Sample:* a subset of the population
- *Parameter:* a numerical measure that describes a characteristic of a population
- *Statistic:* a numerical measure that describes a characteristic of a sample

For example, you may wish to research the average income of residents of your town. Income is the variable of interest; the values you obtain from a survey are the data. The population of interest is the residents of your town, but you will probably survey a small sample of the total population. The average income from the sample is a statistic that you will use to make inferences about the population parameter, the average income of all residents.

Identifying Types of Variables

Variables are of two main types: *categorical* (qualitative) and *numerical* (quantitative).

- Categorical variables typically require answers such as yes/no or some sort of descriptor (blue/brown/green eyes, for example). An easy way of deciding if a variable is categorical, ***you cannot average these responses*** – you can only tally the number of answers for each response.

- Numerical variables will require numerical answers (how many siblings someone has or what your GPA is, for example). Numerical variables can be further defined as discrete or continuous.
  - *Discrete* variables are those that have whole numbers as responses (the number of siblings you have, for example, must be a whole number!).
  - *Continuous* variables are those that can take on any value within a range of values (your GPA can be any decimal number between 0.00 and 4.00 – you don't necessarily need to have a 0, 1, 2, 3, or 4!).

Measurement Scales

There are different ways to measure variables. This helps us order the data and understand whether the differences between values are meaningful (e.g., one value is twice another) or not.

Numerical (quantitative) variables are defined using either an *interval* scale or a *ratio* scale. Data that is measured by an interval scale are ordered, but there is no true zero point. Temperature is a typical example here: when an 80° day is not twice as warm as a 40° day; it's just warmer. Zero degrees does not indicate an absence of heat, so "zero" is an arbitrary point.

A ratio scale is an ordered numerical scale that has a true zero point, and the ratios between values are meaningful. For example, if you have $80 and your friend has $40, you have twice the amount your friend has. If another friend has $0, this means something: there is an absence of money!

Categorical (qualitative) data can be measured on a *nominal* scale or an *ordinal* scale. The word "nominal" comes from the Latin word "nomine," meaning "name." Nominal scales express names of data, but no order is implied. For example, some popular cellular providers are AT&T, T-Mobile, Sprint, and Verizon. If we were to take a survey about cellular providers, the listing of these names is immaterial and do not imply anything about which company is more popular than the others, provides better service, etc. We simply listed them alphabetically.

If data that are measured on an ordinal scale, an ordering or ranking is implied. For example, rating an instructor as poor, fair, good, or excellent implies that "excellent" is a better ranking than "fair." However, there is no way of knowing if the difference between poor and fair is the same as that between good and excellent.

Sampling Data

As mentioned earlier, we have populations of interest we wish to study. This is often accomplished by taking samples from the population. There are several ways to collect data. You could collect data from a published source (e.g., the Statistical Abstracts of the United States), from an experiment, from a survey, or by observing the group of interest in their natural setting. When sampling, you may use either *nonprobability samples* and *probability samples.*

When using nonprobability samples, we don't know the probabilities of selecting individuals. There are two types of nonprobability samples: *judgment samples* and *convenience samples.* When using judgment samples, we use preselected experts and solicit their opinions. While they may be quite knowledgeable on the given topic, their opinions may not be generalized to the population.

Convenience samples ae just that – we collect data in a convenient way. For example, if we were interested in BCC students' opinions about the college cafeteria, we might ask the first 50 students we see outside G Building. While this is an easy, inexpensive way to collect data, we may not get a sample that is representative of the entire BCC student population.

There are several types of probability sampling techniques: *simple random samples, systematic samples, stratified samples,* and *cluster samples.* When using a simple random sample, every item in the population (sometimes called the sampling frame) has an equal probability of being chosen. A random number generator is often used to find the items to include in the sample, so if #17 comes up, we choose the 17th item from the population; if #159 comes up, we choose the 159th item; and so on until our sample is complete.

A systematic sample uses a system to select items from the population. For example, we might start at the 10th item in the population and select every 100th after that, so item #10, #110, #210, etc. would be included in the sample.

When creating a stratified sample, you first subdivide the population into several layers, or strata, such as age groups, family income levels, etc. You then randomly select items from each stratum to include in the sample.

When creating a cluster sample, you subdivide the population into naturally occurring subdivisions such as neighborhoods, counties, or sales territories. You then randomly select items from one or more clusters to include in your sample.

The most important thing to remember about sampling is that your sample should be a miniature version of the population.

Time Series vs. Cross-sectional Data

Quantitative data can be cross-sectional or time series. Most of the data you will be seeing in this course will be cross-sectional: these values are collected from a sample at a given point in time. Time series data are data covering a specific situation or topic (e.g., corporate sales) over a relatively long period of time.

Elaine A. Previte

Organizing & Visualizing Data
Chapter 2 Summary

Data Collection

Data can be collected in several ways. You can use either a *primary data source* (such as your own survey) or a *secondary data source* (such as data from other sources such as the government or an industry association).

Organizing Categorical Data

There are several ways to organize categorical (qualitative) data. A *summary table* consists of a listing of each category along with the number or percentage of responses of total responses. A *contingency table* is similar, but also further breaks down that data by also including a second variable. See the hypothetical sample of students reported below. The contingency table takes the data from the summary table and refines it to include information about choice of major by gender, and thus adds a second "dimension," so to speak.

Summary Table	
Major	***Number***
Accounting	35
Finance	20
Management	30
Marketing	15
Total	100

Contingency Table			
Major	***Male***	***Female***	***Total***
Accounting	15	20	35
Finance	15	5	20
Management	20	10	30
Marketing	5	10	15
Total	55	45	100

Organizing Numerical Data

Numerical (quantitative) data can also be arranged in several ways. The first way is to use an *ordered array* (a list that ranks data from the smallest to the largest). This may be useful for small amounts of data, but can be cumbersome when dealing with large data sets.

A *frequency distribution* tallies the data and summarizes them by placing them into *classes (e.g., 10-19, 20-29, 30-39, etc.).* For example, let's say we have 50 students ranging in age from 20 to 35. Placing them into the classes mentioned above would provide too little information, and listing all 50 pieces of data would be too cumbersome, so a rule of thumb is to have between 5 and 15 classes. So, for this example, we would calculate the *class width* by taking the range of the data (35 – 20 = 15, the range) and dividing by the number of classes desired. The class width is the difference between the first value of one class and the first value of the next class. If we wish to have 6 classes, the class width would be 15/6 = 2.5 (round up to 3.0). Our classes would be as follows:

Class	Frequency
20-22	22
23-25	15
26-28	5
29-31	2
32-34	4
35-37	2
Total	50

A frequency distribution allows you to make observations about how the data are distributed.

Other terms that are used when discussing frequency distributions are *class limits* and *class midpoints. Class limits* are the two numbers that define the beginning and end of each class; for example, 20 is the lower limit of the first class above, and 22 is the upper limit. *Class midpoints* are just that: the number that is in the middle of each class. They ay be found by taking the average of the lower and upper limits of each class. The class midpoint for the first class is (20 + 22)/2 = 42/2 = 21. And since our class width is 3, the difference between successive class midpoints will also be 3. So, the next midpoint will be 24, the next is 27, etc. See the table below.

A frequency distribution may be modified in a number of ways. For example, we could find the *relative frequency* of each class or the *percentage* of values in each class. The *relative frequency* is found by taking the class frequency and dividing by the total number of pieces of data in the data set. The percentage simply translates that decimal value to a percentage.

Class Limits	Class Midpoint	Frequency	Relative Frequency	Percentage (%)
20-22	21	22	22/50 = 0.44	44
23-25	24	15	15/50 = 0.30	30
26-28	27	5	5/50 = 0.10	10
29-31	30	2	2/50 = 0.04	4
32-34	33	4	4/50 = 0.08	8
35-37	36	2	2/50 = 0.04	4
Total		50	1.00	100

Cumulative percentage distributions show the percentage of the data that lie below a specified value. For example, if you need to know the percentage of students who are younger than 26, you would add 44% + 30% = 74%. The cumulative percentage distribution shows the cumulative percentage of data that are less than the lower class limit of a particular class. Your last class will always have a cumulative percentage of 100%.

Class	Percentage (%)	Cumulative Percentage
20-22	44	44
23-25	30	74
26-28	10	84
29-31	4	88
32-34	8	96
35-37	4	100
Total	100	

Visualizing Categorical Data

As you may know, categorical data can be represented in several ways. The bar chart compares different categories by using bars to represent the total number of values in each category. Bar charts may be drawn with either a horizontal or vertical orientation.

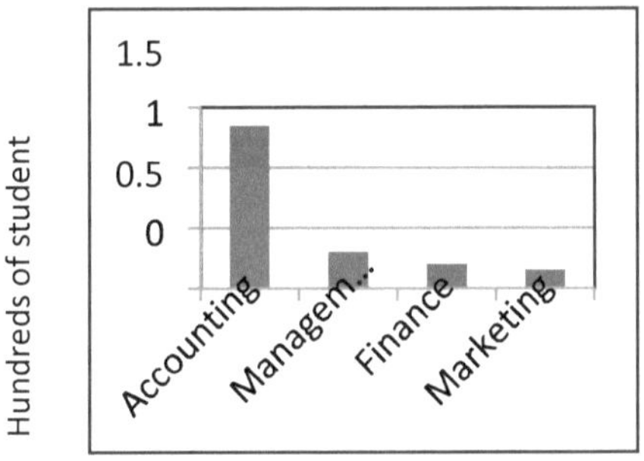

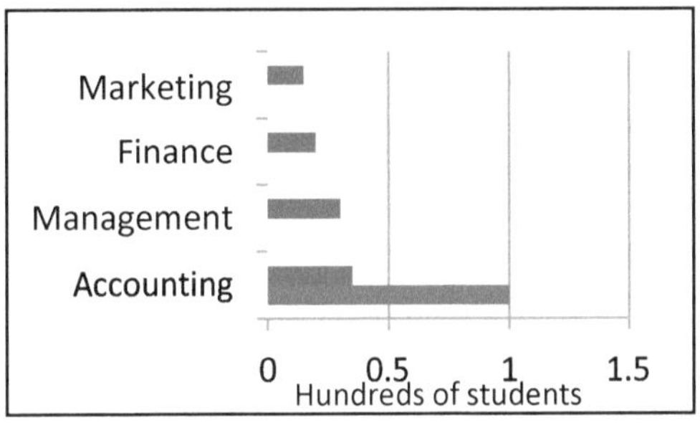

Side-by-side bar charts are used to show joint responses from two categorical variables.

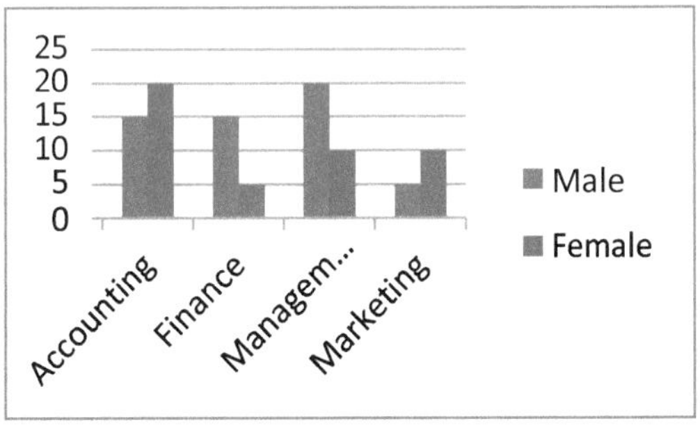

A *pie chart* uses sectors of a circle to represent the tally of each category. To find the size of the central angle for a particular sector, take the relative frequency of the category and multiply by 360 degrees.

Major	***Number***	***Central Angle (degrees)***
Accounting	35	.35 x 360 = 126.0
Finance	20	.20 x 360 = 72.0
Management	30	.30 x 360 = 108.0
Marketing	15	.15 x 360 = 54.0
Total	100	360.0

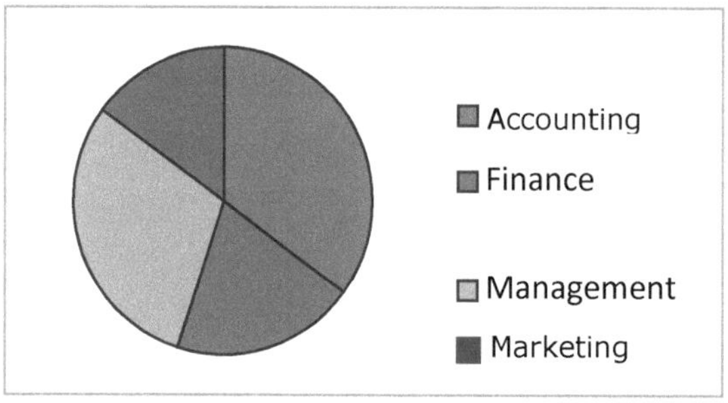

A Pareto Chart shows the tallies for each category along with cumulative percentages. The categories are organized from largest percentage down to the smallest. The line graph always increases (or is flat); it never decreases as this represents cumulative values.

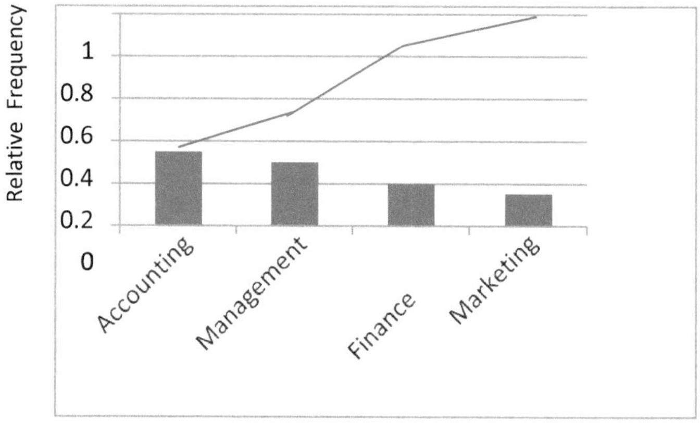

Elaine A. Previte
May 2019

Visualizing Numerical Data

A *stem-and-leaf display* allows you to see how data are distributed and where areas of concentration exist. Let's say the following data represent exam scores for the first statistics exam last semester. The tens digits go in to stem and the ones digits are the leaves.

45, 56, 67, 67, 68, 75, 76, 78, 78, 80, 82, 82, 84, 85, 85, 92, 100

Stem	Leaves
4	5
5	6
6	7 7 8
7	5 6 8 8
8	0 2 2 4 5 5
9	2
10	0

A *histogram* is a bar chart for grouped numerical data. The midpoint for each class is plotted on the horizontal axis. This is different from a bar chart in that the bars touch (since the classes are adjacent to each other).

Class	Percentage (%)
20-22	44
23-25	30
26-28	10
29-31	4
32-34	8
35-37	4
38-40	0
Total	100

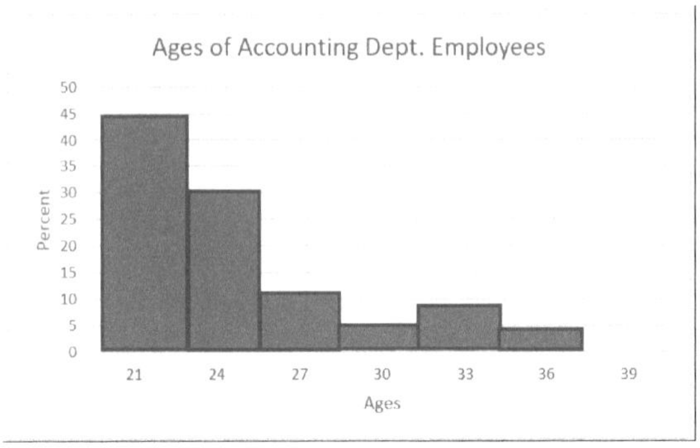

Histograms often exhibit different shapes. These shapes give us insight into the distribution of the actual data values.

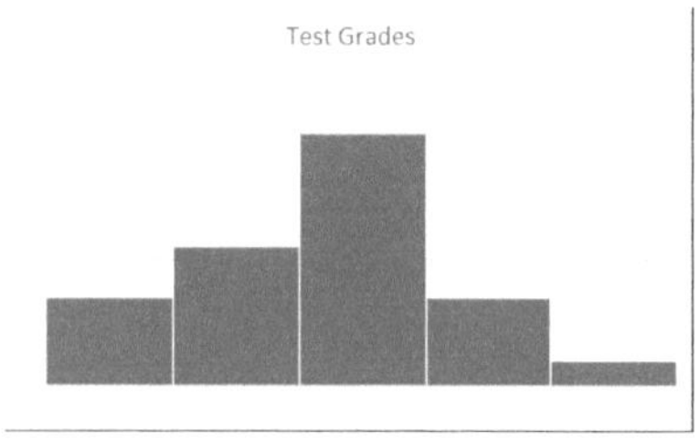

This histogram is roughly symmetric; that is, it is "mounded" and "even" relative to the center of the graph.

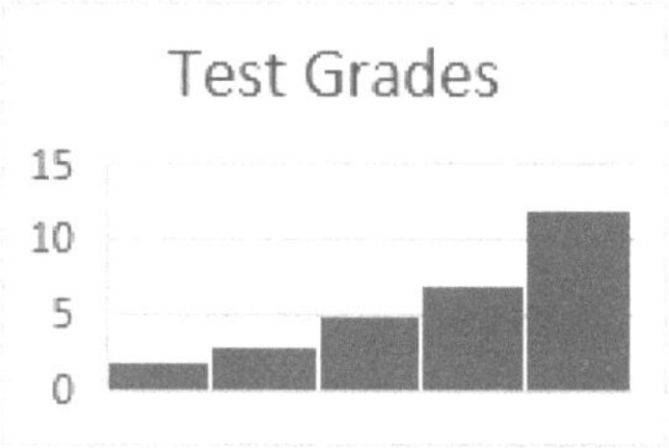

This histogram is skewed left; there are a few lower data values and many more pieces of data on the right-hand side.

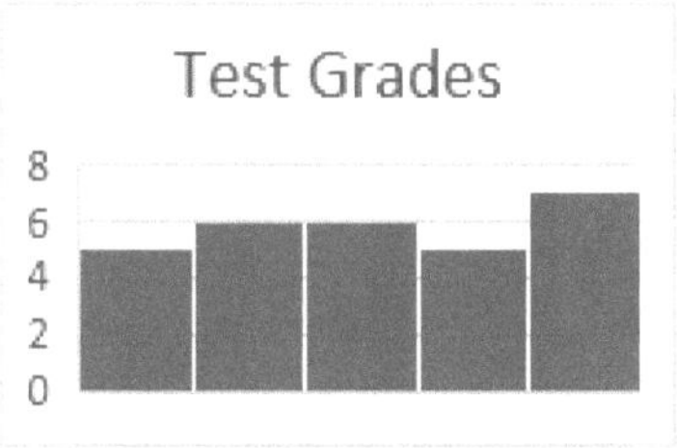

This histogram is roughly uniform; there are approximately the same number of pieces of data in each class.

A *percentage polygon* connects the midpoints of each class with a line graph. The graph begins and ends on the horizontal axis in order to create a closed figure (which is what a polygon is by definition).

A *cumulative percentage polygon,* or *ogive (pronounced "oh-jive"),* uses the data from the cumulative percentage distribution. The lower limit of each class is the x-coordinate, and the cumulative percentage is the y-coordinate. It is important to note that, since this graph shows the cumulative data, it will always grow (or, at the least, flatten out) as you move from class to class. You will always end up at 100%. (See next page.)

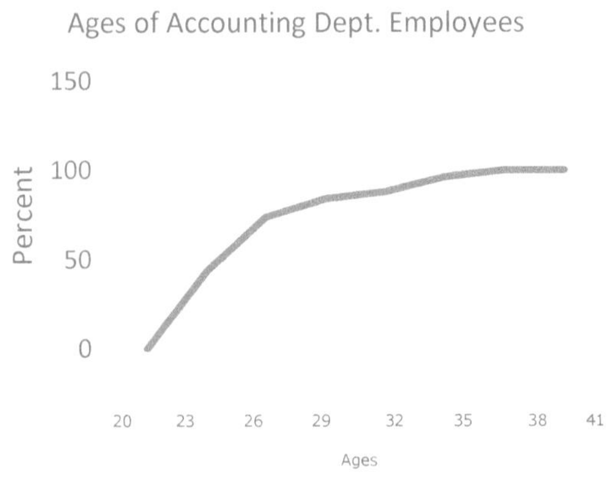

Visualizing Two Numerical Variables

A *scatter plot* shows a plot of data that consists of two numerical values (e.g., height vs. weight). The scatter plot below shows apartment size (in square feet) vs. monthly rent (in dollars).

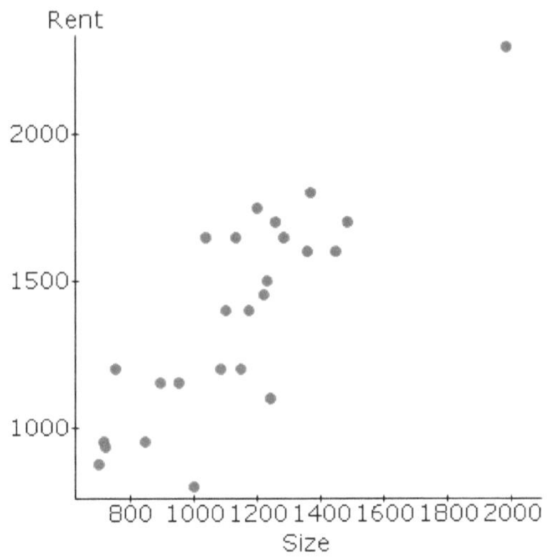

A *time-series plot* is a scatter plot that has time as the horizontal variable. The time-series plot below shows gasoline prices vs. time from January 2006 to June 2010.

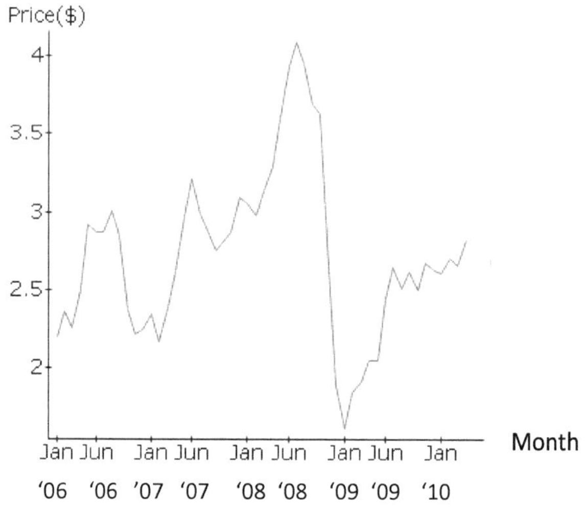

Elaine A. Previte
May 2019

Misuses and Common Errors in Visualizing Data

Some common mistakes that are made when illustrating data, including:

- Omitting labels from axes/No zero point on the graph

Example: Sometimes you'll see graphs like the one below featured in TV commercials or magazine ads. Because there are no labels on the vertical axis, it appears as though Brand B contains many times the pain reliever that is found in Brand A. However, by including the zero and labeling the axes properly, we can see that, while Brand B contains more pain reliever, it contains only twice the amount in Brand A, not six times that amount.

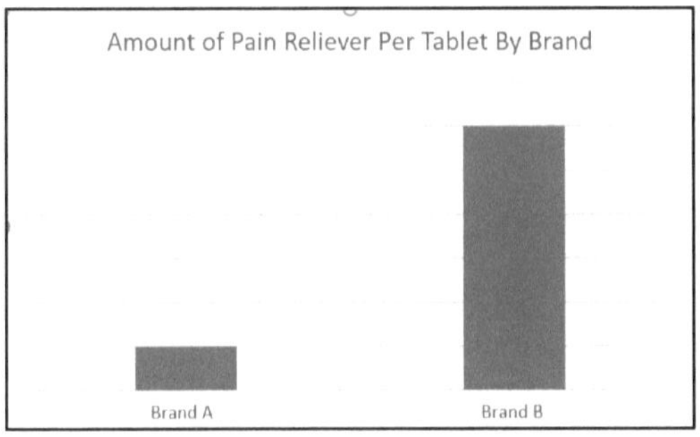

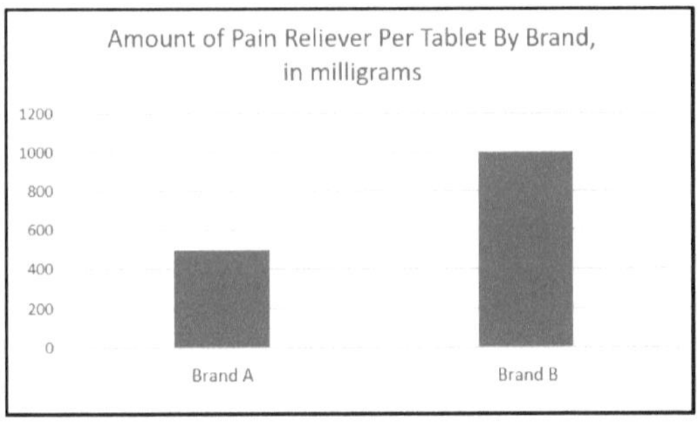

- Showing disproportionate graphics

Publications often include amusing graphics to illustrate statistical data. However, sometimes those graphics may be misleading. For example, in the table below, the housing prices in 2000 are 6 times those in 1950, but the 2000 house looks like much more than 6 times the area of the 1950 house. It has been stretched to increase both width and height, so the area is well over six times the 1950 house. This gives the impression that the 2000 house is several orders of magnitude larger than the 1950 house, and is thus misleading.

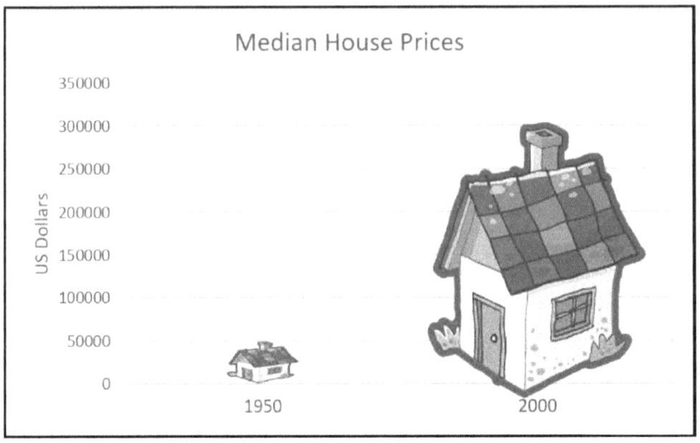

Calculating Descriptive Statistics
Chapter 3 Summary

Measures of Central Tendency

Measures of *central tendency* give information about the "center" of the data.

There are several measures of central tendency:

- The mean (or, arithmetic mean – pronounced ar-ith-**MET**-ic)
- The median
- The mode
- The geometric mean (we will omit this topic)
- The weighted mean

The *arithmetic mean* is most commonly called the mean, or the average, of a data set. When referring to the mean of a sample, we use the symbol $\bar{x}$ (read "x-bar"). You already know how to compute it:

$$\bar{x} = \frac{sum\ of\ all\ the\ values}{number\ of\ values} = \frac{x_1 + x_{2+} \cdots + x_n}{n}$$

We use a capital Greek letter, $\sum$ (pronounced "sigma"), to indicate that we wish to add (to sum) some values. The subscripts on the x's above are there to differentiate among them, and in summation notation the subscript ***i*** is called the index. The index begins at 1 and continues to the number of the last piece of data, ***n*** (***i*** = 1, 2, 3, …, n). So, the definition of mean can be written as follows:

$$\bar{x} = \frac{\sum_1^n x_i}{n} = \frac{x_1 + x_{2+} \cdots + x_n}{n}$$

Here is an example: suppose you have 5 test grades in a course. What is your average?

5 test grades: 83, 75, 76, 98, 100 (n = 5 pieces of data)

$$\bar{x} = \frac{\sum_1^5 x_5}{5} = \frac{83 + 75 + 76 + 98 + 100}{5} = \frac{432}{5} = 86.4\ average$$

Elaine A. Previte
May 2019

Again, the small numbers written next to the $\sum$ are called the indices of the summation; in other words, the first term to be added is x_1 and the last one is x_n (n will be specified in any given problem; in our example above, n was 5).

End at the nth piece of data, x_n

$$\sum_1^n x_i \leftarrow \text{index}$$

Start the summation at the first piece of data

Note: After a few examples we will dispense with the indices – they just seem to add clutter to our formulae!

The *median* is the middlemost piece of data when the data are ordered lowest to highest. The position of the median can be found as follows:

$$position\ of\ median = \frac{n+1}{2}\ (rank\ of\ value).$$

Again, note that this will give you the *position* of the median, *not* its value.

You can compute the median as follows:

- If the data set has an odd number of pieces of data, the median is the middle value.
- If the data set has an even number of pieces of data, the median will be the average of the middle two values.

For example:

- 0, 15, 16, 17, 18, 20, 100 ➔ 7 pieces of data; the median is located at the $\frac{7+1}{2} = 4th$ position, so the median is 17.

- 0, 15, 16, 17, 18, 20, 100, 3000 ➔ 8 pieces of data; the median is located at the $\frac{8+1}{2} = 4.5th$ position (the average of the 4th and 5th pieces of data), so the median is $\frac{17+18}{2} = 17.5$.

Notice that the median is not affected by extremely small or large values – the median simply reflects the value at the center of the data set.

The *mode* is the value in the data set with the highest frequency. A data set can have one mode (unimodal), two modes (bimodal), several modes, or no mode.

For example:

- 3, 3, 5, 5, 5, 6, 6, 6, 7, 8, 9, 10 → the mode is 5 since its frequency is highest (its frequency is 3).
- 3, 3, 5, 5, 5, 6, 6, 6, 7, 8, 8, 8 → this data is *bimodal* since both 5 and 8 have frequency 3.
- 1, 2, 3, 4, 5, 6, 7, 8, 9, 10 → this data has no mode as all data have frequency 1.

A *weighted mean* is an average that does not weigh each piece of data equally. The most familiar example of a weighted mean is your grade point average (GPA). You know that an A in CSS 101 doesn't "count" as much toward your GPA as does an A in ACC 101! The course grade is weighted by number of credits and yields the "quality points" that are used in the GPA calculation.

For example:

Course	**Credits**	**Grade**	**Quality Points (#Credits x Grade)**
CSS 101	1	A = 4.0	1 x 4 = 4.0
ACC 101	4	A = 4.0	4 x 4 = 16.0
ECN 111	3	D = 1.0	3 x 1 = 3.0
MAN 101	3	B = 3.0	3 x 3 = 9.0
MAR 101	3	C = 2.0	3 x 2 = 6.0
Totals	14		38.0

$$GPA = \frac{Total\ number\ of\ Quality\ Points}{Total\ number\ of\ Credits} = \frac{38}{14} = 2.714$$

Mean of a Frequency Distribution

If your data have already been put into classes and are in a frequency table (i.e., you don't have the original data), you can still find the mean of the data set. It will probably not be the same as the mean of the original data set, but it should be close to that value.

Let's use the age data from the last chapter. Since we don't know the exact values that are contained in each class, we need a stand-in: the class midpoint will do this for us. Then we will take the product of each midpoint and its class frequency; this will give us an approximation for the sum of all the pieces of data in each class.

Class	Class Midpoint, m_i	Frequency, f_i	$m_i \cdot f_i$
20-22	21	22	$21 \cdot 22 = 462$
23-25	24	15	$24 \cdot 15 = 360$
26-28	27	5	$27 \cdot 5 = 135$
29-31	30	2	$30 \cdot 2 = 60$
32-34	33	4	$33 \cdot 4 = 132$
35-37	36	2	$36 \cdot 2 = 72$
Total		50	1221

The mean of this frequency distribution is $\bar{x} = \frac{1221}{50} = 24.42$.

The formula is: $\bar{x} = \frac{\Sigma m_i \cdot f_i}{n}$ $or\ \bar{x} = \frac{\Sigma m_i \cdot f_i}{\Sigma f_i}$. Note that we divide by the total number of pieces of data (50 in this case) and ***not*** by the total number of classes. Also note that we always follow the order of operations when computing any statistic: i.e., we multiply before we add and we compute the sum in the numerator before we divide by the value in the denominator.

Variation and Shape

Variation measures the amount of dispersion, or spread, of the data away from the central value. *Shape* refers to the pattern of the distribution of the data.

There are several ways to discuss the variation of a set of data.

- *Range:* the difference between the highest and lowest values
- *Variance:* an "average" of the sum of squared deviations from the mean
- *Standard Deviation:* the square root of the variance
- *Covariance: measures the type of relationship between two variables (positive, zero, or negative)*
- *Coefficient of Variation:* the ratio of the standard deviation to the mean of a data set
- *z-score:* a measure of position (i.e., the number of standard deviations a data value is from the mean of the data set)

The range is the easiest measure of variation to calculate. It is simply the highest data value minus the lowest:

$$Range = Highest\ value - Lowest\ Value.$$

Given the data set of five test grades above (75, 76, 83, 98, 100), we find the range by subtraction:

$$Range = 100 - 75 = 25$$

The *variance* is a very important measure of dispersion. The sample variance is symbolized by s^2, and is a kind of average of the squared deviations (differences) of each value from the mean.

$$s^2 = \frac{\sum_1^n (x_i - \bar{x})^2}{n-1}$$

Let's use the same data set as above to compute the sample variance. It is important to remember that you need to follow the order of operations when calculating the variance. Recall $\bar{x} = 86.4$.

Data, x_i	$\boldsymbol{x_n - \bar{x}}$	$\boldsymbol{(x_i - \bar{x})^2}$
75	75 – 86.4 = -11.4	$(-11.4)^2$ = 129.96
76	76 – 86.4 = -10.4	$(-10.4)^2$ = 108.16
83	83 – 86.4 = -3.4	$(-3.4)^2$ = 11.56
98	98 – 86.4 = 11.6	$(11.6)^2$ = 134.56
100	100 – 86.4 = 13.6	$(13.6)^2$ = 184.96
$\sum x = 432$		$\sum(x_i - \bar{x})^2$ = 569.2

So, the variance is:

$$s^2 = \frac{\sum_1^n (x_i - \bar{x})^2}{n-1} = \frac{569.2}{5-1} = \frac{569.2}{4} = 142.3$$

Because we square the differences when calculating the variance, we really end up also squaring the units, so the variance calculated above is 142.3 "square points" (which is pretty meaningless!). Consequently, we use another measure of variation, the *standard deviation,* which will get us back to "regular" units since it is the square root of the variance. The symbol for sample standard deviation is *s*.

$$s = \sqrt{s^2} = \sqrt{\frac{\sum_1^n (x_i - \bar{x})^2}{n-1}}$$

Hence the standard deviation for the data above is

$$s = \sqrt{s^2} = \sqrt{\frac{\sum_1^n (x_n - \bar{x})^2}{n-1}} = \sqrt{\frac{569.2}{4}} = \sqrt{142.3} = 11.92896\ points.$$

Some texts provide the so-called "short-cut" formula for computing sample variance and standard deviation, and this version is often easier for computational purposes. Of course, you will get the same result regardless of which formula is used.

$$s = \sqrt{\frac{n\sum x^2 - (\sum x)^2}{n(n-1)}}$$

For our example, $\sum x = 432\ and\ \sum x^2$=37894, and n = 5, so

$$s = \sqrt{\frac{n\sum x^2 - (\sum x)^2}{n(n-1)}} = \sqrt{\frac{5(37894)-(432)^2}{5(5-1)}} = \sqrt{\frac{189470-186624}{5(4)}} = \sqrt{\frac{2846}{20}} = \sqrt{142.3} = 11.92896.$$

Voila!

Interpreting Standard Deviation

In and of itself, standard deviation is meaningless. Standard deviation does have meaning when comparing the spread in two or more data sets (e.g., our in-class group work comparing three data sets with the same mean). For example, a data set with a standard deviation of 4.5 is relatively more dispersed ("spread out") than a data set with a standard deviation of 2.5.

We can also use some well-known rules to help us interpret standard deviation. The first is called Chebyshev's Theorem and applies to distributions of *any* shape. Chebyshev states that one should see *at least* $1 - \frac{1}{k^2}$ of data within *k* standard deviations of the mean (k ≠1). For example, if looking within 2 standard deviation of the mean of a data set, you should see *at least* $1 - \frac{1}{k^2} = 1 - \frac{1}{2^2} = 1 - \frac{1}{4} = \frac{3}{4}$ or 75 % of the data. If looking within 3 standard deviations, you should see *at least* $1 - \frac{1}{k^2} = 1 - \frac{1}{3^2} = 1 - \frac{1}{9} = \frac{8}{9}$ or 88.89 % of the data. The reason Chebyshev says "at least" is because this theorem applies to data of any shape, so he is cautious about over-estimating the percentages.

As an example, let's assume that the price of houses sold in Bristol County during the past 30 days averaged $305,000 with a standard deviation of $40,000. Since we don't know the shape of the distribution of house prices, we will rely on Chebyshev's Theorem to provide analysis. What percent of homes sold in the $225,000 – $385,000 range? Since the mean price was $305,000, we can see that $225,000 is two standard deviations below the mean (305.000 – 225,000 = 80,000; 80,000 ÷ 40,000 = 2; likewise, 385,000 – 305,000 = 80,000; 80,000 ÷ 40,000 = 2). Therefore, by Chebyshev, at least $1 - \frac{1}{k^2} = 1 - \frac{1}{2^2} = 1 - \frac{1}{4} = \frac{3}{4}$ or 75% of homes in Bristol County had selling prices between \$225,000 and \$385,000.

Another useful tool is called the Empirical Rule. This rule applies only to data that are mounded, or bell-shaped (also called "normal"). You must know that your data are bell-shaped (see below) in order to utilize the Empirical Rule.

The Empirical Rule states that, given a bell-shaped distribution, you should see approximately 68% of the data within one standard deviation of the mean of the distribution ("within one standard deviation" means from one standard deviation below the mean to one standard deviation above the mean). Approximately 95% of the data lie within two standard deviations of the mean, and approximately 99.7% lie within 3 standard deviations. That means that 0.3% of the data lie outside 3 standard deviations, and since the curve is symmetric about the mean, then 0.15% of the data lie below -3 standard deviations and above +3 standard deviations (note: 0.15% means 15/10,000).

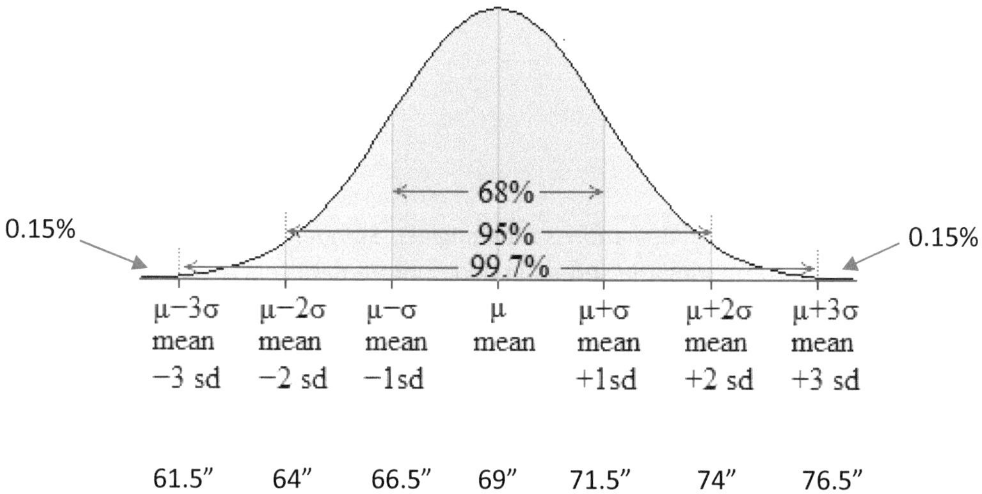

Men's Heights

For example, the mean height of college-aged men is 69 inches, with a standard deviation of 2.5 inches. That means that approximately 68% of college-aged men are between 66.5 and 71.5 inches; approximately 95% are between 64 and 74 inches; and approximately 99.7% are between 61.5 and 76.5 inches. 0.15% are shorter than 61.5 inches, and 0.15% are taller than 76.5 inches (so not too many men would be eligible for the NBA!).

The importance of the Empirical Rule is the universality of its application. Many phenomena in life have distributions that are bell-shaped: heights, weights, dress sizes, shoe sizes, SAT scores, IQ scores, etc. The beauty of the Empirical Rule is that the percentages shown above apply to all these situations (and more!). So, for example, if the average American woman's dress size is 16 (yes -size 16!) with a standard deviation of 4, then approximately 68% of American women are between sizes 12 and 20;

approximately 95% are between sizes 8 and 24; and approximately 99.7% are between sizes 4 and 28. Only 15 in 10,000 women wear smaller than a size 4; 15 in 10,000 wear larger than a size 28.

Exploring Numerical Data

Data can also be examined by looking at the position of each piece of data, or certain key positions. *Quartiles* split the data into four parts (i.e., four quarters). The first quartile is located at this position for Q_1 : $\frac{n+1}{4}$. Note: the data *must be ordered* before locating quartiles, etc.

Let's use this set of data to illustrate:

5, 29, 36, 42, 55, 58, 67, 68, 75, 80 (n = 10 pieces of data).
1 2 3 4 5 6 7 8 9 10

So, $Q_1 = \frac{n+1}{4} = \frac{10+1}{4} = \frac{11}{4} = 2.75$. This means that the location of Q_1 is the 2.75th piece of data, but we round up to the 3rd piece of data, so Q_1 =36. This means that 25% of the data are smaller than 36; 75% are larger.

Q_2 is the median of the data. Since there are 10 values, the median is located at the 5.5th piece of data (i.e., the average of the 5th and 6th values). Hence, Q_2 is $\frac{55+58}{2} = 56.5$. 50% of the data are smaller than this value; 50% are larger.

Q_3 is the third quartile, so it is located three-quarters of the way from the smallest value to the largest. $Q_3 = \frac{3(n+1)}{4} = \frac{3(10+1)}{4} = \frac{33}{4} = 8.25 \rightarrow 8$. So, the third quartile is located at the 8th position: Q_3 = 68. 75% of the data are smaller than Q_3; 25% are larger.

The *Interquartile Range* is the difference between Q_3 and Q_1:
$IQR = Q_3 - Q_1$, so in this case the IQR = 68 – 36 = 32. This means that the middle 50% lie between the values 36 and 68.

The *Five-Number Summary* consists of the following values: Lowest value, Q_1, median, Q_3, and the largest value. In this example, the five-number summary is: 5, 36, 56.6, 68, and 80. Recall we saw the five-number summary when we did one-variable stats on the TI-84.

A *boxplot* (or *box-and-whiskers display*) is a visualization of the five-number summary. You can tell if the data are symmetric or skewed by analyzing the boxplot. From looking at the boxplot below, you can see that the data are skewed left.

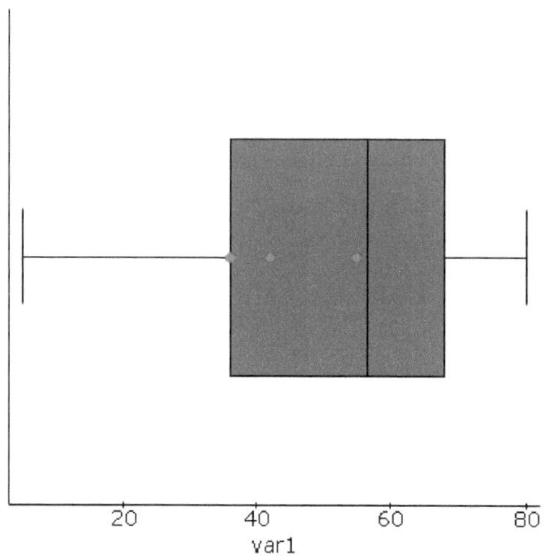

Numerical Descriptive Measures for a Population

We can calculate descriptive values for a population. When doing so, we use Greek letters to symbolize these *population parameters*.

The population mean is μ (pronounced "mu" or "mew"). The calculation is identical to that for the sample mean.

$$\mu = \frac{sum\ of\ all\ the\ values}{total\ number\ of\ values} = \frac{\sum x_i}{N} = \frac{x_1 + x_{2+} \cdots + x_N}{N},$$

where N is the number of values in the population. Note that it is important to differentiate between lower-case n (sample size) and upper-case N (population size).

If we are finding the variance or standard deviation of a population of data (not a sample taken from the population), we must modify the formulae above. The population variance is called σ^2, read "sigma squared." This is a lower-case sigma, not to be confused with capital sigma, Σ, which indicates summation. And since the standard deviation is the square root of the variance, the population standard deviation is sigma $\sigma = \sqrt{\sigma^2}$. The formulae are as follows:

$$\sigma^2 = \frac{\sum(x-\mu)^2}{N} \text{ and } \sigma = \sqrt{\sigma^2} = \sqrt{\frac{\sum(x-\mu)^2}{N}}.$$

Elaine A. Previte
May 2019

The Covariance and the Coefficient of Correlation

In the case where we have two related variables (e.g., x and y) we can measure how well the two are related. The covariance is one way of determining this relationship. A positive covariance means that there is a positive relationship between the variables (i.e., a positive slope when graphed in the coordinate plane), and a negative covariance means there is a negative relationship (negative slope). A covariance of zero means there is no relationship between the variables.

For example, let's look at the following data (no story here; just calculations!). Note that $\bar{x} = \frac{122}{4} = 30.5, and\ \bar{y} = \frac{440}{4} = 110.$ The formula for the covariance is: $COV(x, y) = \frac{\Sigma(x-\bar{x})(y-\bar{y})}{n-1}$. ***Remember to follow the order of operations.***

x	y	$x - \bar{x}$	$y - \bar{y}$	$(x - \bar{x})(y - \bar{y})$
21	80	21 – 30.5 = -9.5	80 – 110 = -30	(-9.5)(-30)= 285
25	120	25 – 30.5 = -5.5	120 – 110 = 10	(-5.5)(10) = -55
40	140	40 – 30.5 = 9.5	140 – 110 = 30	(9.5)(30) = 285
36	100	36 – 30.5 = 5.5	100 – 110 = -10	(5.5)(-10) = -55
122	440			460

Note that the sum in the bottom of the last column gives us the numerator of the covariance and, since n = 4 ordered pairs, n – 1 = 3. Thus, the covariance is:

$COV(x, y) = \frac{\Sigma(x-\bar{x})(y-\bar{y})}{n-1} = \frac{460}{3} = 153.33.$ We have no idea if this value represents a "good" or "bad" covariance; all we know is that it is a positive value, meaning that these ordered pairs would have a positive slope when graphed.

The correlation coefficient, represented by the letter ***r***, brings the covariance "down to size" by dividing it by the product of the standard deviations of x and y. We won't compute the standard deviations here; just trust that they are as follows: $s_x = 8.963\ and\ s_y = 25.820.$ Consequently,

$$r = \frac{COV(x,y)}{s_x \cdot s_y} = \frac{153.33}{(8.963)(25.820)} = \frac{153.33}{231.425} = 0.6626.$$

How do we interpret this result? Well, a perfect negative correlation will result in an r of negative one, while a perfect positive correlation will result in an r of positive one. If there is little or no relationship between the two variables, then r will be very close to zero.

-1	0	+1
perfect negative correlation	no correlation	perfect positive correlation

Elaine A. Previte
May 2019

Here are some examples of graphs with their associated correlation coefficients.

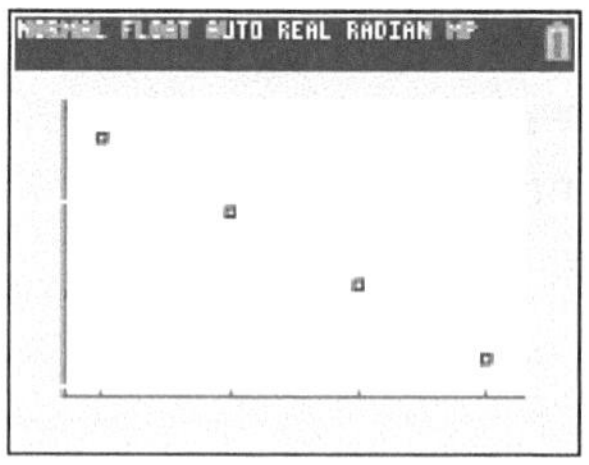

r = -1

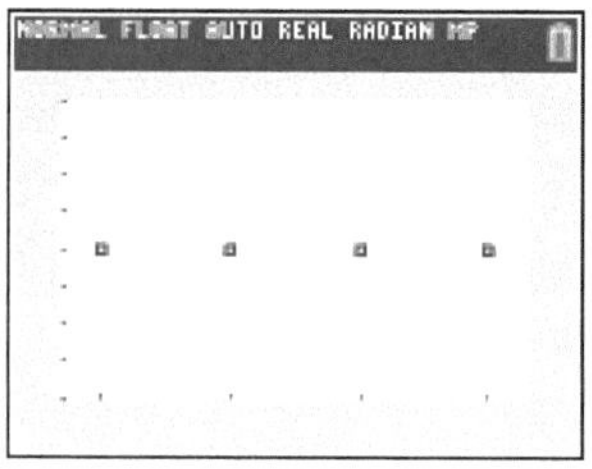

r = 0

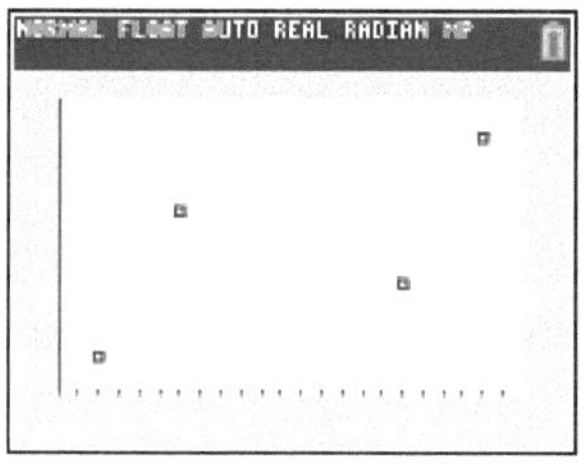

r = 0.6626
(our data above)

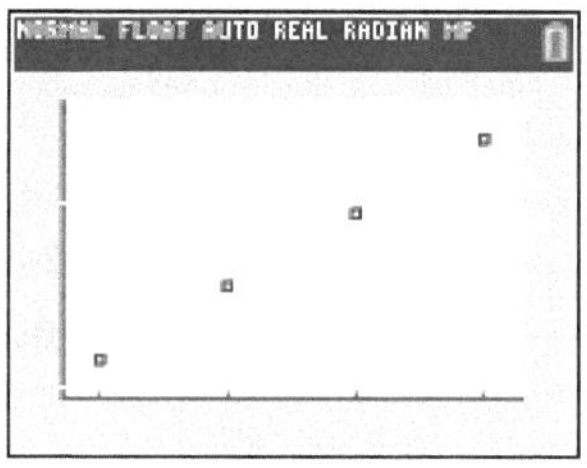

r = +1

z-scores

A way of measuring the position of a piece of data is by calculating its z-score. A z-score tells us the number of standard deviations away from the mean for that particular piece of data. For example, SAT scores have a mean of 500 and a standard deviation of 100. If you were to score a 600 on that test, you would be 100 points above the mean and, since a standard deviation is 100 points, your z-score would be +1. If a friend scored 400 on the same test, his/her z-score would be -1 since he/she scored 100 points below the mean.

The formula for the z-score is $z = \frac{x-\bar{x}}{s}$ or $z = \frac{x-\mu}{\sigma}$. Which version of the formula you use is determined by whether you have sample data or population data. But the important thing to notice is that z-scores are related to the Empirical Rule mentioned above. The z-score simply measures how many standard deviations away from the mean a particular piece of data lies. However, a z-score need not be an integer: i.e., a piece of data may lie 1.25 standard deviations above the mean or 0.7 standard deviations below it.

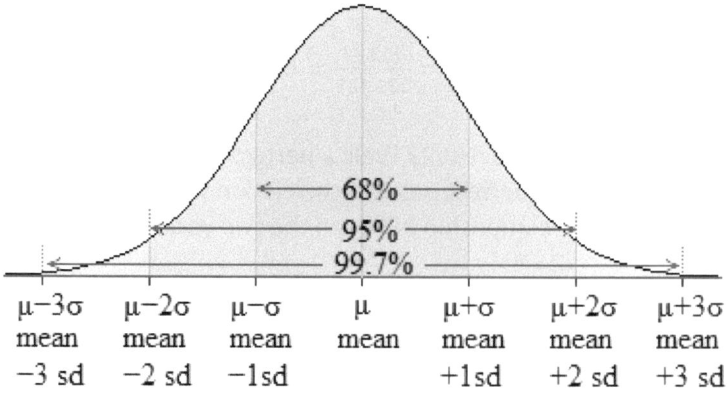

z-scores: **-3** **-2** **-1** **0** **+1** **+2** **+3**

Elaine A. Previte
May 2019

Basic Probability
Chapter 4 Summary

Basic Probability Concepts

There are three types of probability:

- *A priori*
- Empirical
- Subjective

In *a priori* probability, the probability of an event occurring is determined by prior knowledge you possess. For example, if you were to roll a fair six-sided die, you know there is one way to obtain a 4 out of the 6 possible outcomes. P(4) means the probability of rolling a 4.

$$P(4) = \frac{\text{the number of ways the event can occur}}{\text{total number of outcomes}} = \frac{1}{6}$$

Empirical probability is based on data sets. So the probabilities of certain events will vary as the data sets vary. For example, let's say that 10% of BCC students are nursing majors.
Then the probability of searching the BCC student database and selecting a nursing major is 10%. But the probability of searching the Bryant University student database and selecting a nursing major is 0% as Bryant doesn't offer a nursing program. Empirical probabilities vary from situation to situation, but *a priori* probabilities do not. However, empirical probabilities are computed just as a priori ones are:

$$P(A) = \frac{\text{the number of ways the event A can occur}}{\text{total number of outcomes}}.$$

Subjective probability can be called "guesstimation" as the probabilities are based on one's experience and opinion. For example, you may love the Red Sox, so you might say the probability of their winning the World Series is 60%; a Yankees fan may place a different probability on that event!!

Other vocabulary you need to know:

- Probability Experiment – a chance process that leads to results, called outcomes.
- Event – one of the possible outcomes of a probability experiment.
- Simple event – can occur only one way.

- Joint event – an event that has more than one characteristic and can happen in more than one way.
- Complement of an event – the exact opposite of an event A, denoted A′. For example, the complement of tossing a coin and getting "heads" is getting "tails." (Note: some texts use A^c or $\bar{A}$ to denote the complement of an event, A.)
- Sample space – the set of all possible outcomes of a probability experiment.

For example, let's say we will roll a single (fair) six-sided die. The probability experiment is the roll of the die; the outcomes are the possible number of pips that land face-up; the sample space consists of all these outcomes: S = {1, 2, 3, 4, 5, 6}.

If we were to roll two fair six-sided die, the sample space would be a bit more complicated. The probability experiment is the roll of the dice; the outcomes are the number of pips that land face-up on each of the two die; the sample space would best be represented in tabular form:

Die 2

Die 1

1,1	2,1	3,1	4,1	5,1	6,1
1,2	2,2	3,2	4,2	5,2	6,2
1,3	2,3	3,3	4,3	5,3	6,3
1,4	3,4	3,4	4,4	5,4	6,4
1,5	2,5	3,5	4,5	5,5	6,5
1,6	2,6	3,5	4,6	5,6	6,6

Sample Space for the experiment - tossing two dice

As another example, let's say we will draw one card from a traditional deck (4 suits, 13 cards per suit). The probability experiment is the draw of the card; the outcomes are all 52 cards in the deck; the sample space consists of all 52 cards. See below.

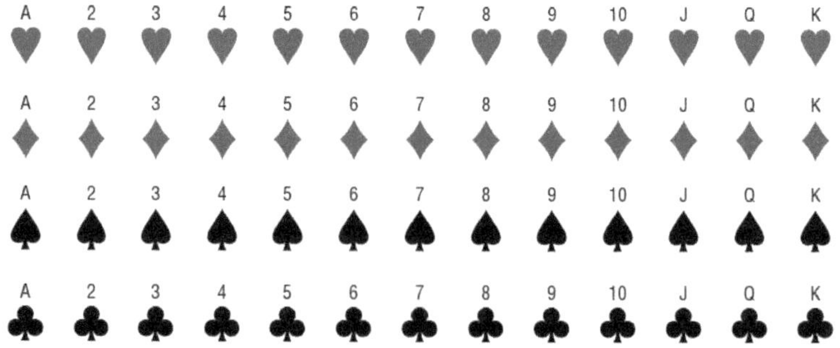

Sample space for a standard deck of cards

If a couple were interested in having three children, the sample space would consist of all the possible combinations of genders: S = {BBB, BBG, BGB, GBB, BGG, GBG, GGB, GGG}.

Another way of illustrating the possible genders of the three children is through the use of a tree diagram:

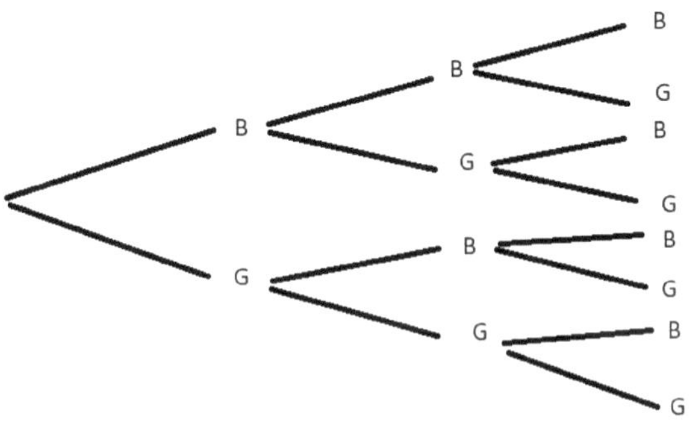

Outcomes:

BBB
BBG
BGB
BGG
GBB
GBG
GGB
GGG

Of course, there are rules for probability. The first few are simply common sense:

1. The probability of an event, A, is between 0 and 1, inclusive; i.e., $0 \leq P(A) \leq 1$.
2. If an event cannot occur, its probability is 0.
3. If an event is certain to occur, its probability is 1.
4. The sum of the probabilities of all the simple outcomes of a probability experiment is always 1. For example, when rolling a fair die the simple outcomes are obtaining a 1, 2, 3, 4, 5, or 6. Since these events are equally likely (i.e., they all have probability 1/6), the sum of the probabilities of these outcomes is 1.
5. The probability of the complement of an event, A, is P(A') = 1 – P(A). For example, if rolling a single die, let's say we're interested in rolling a 4. Let's call that event A. The probability of rolling a 4 is 1/6 (i.e., P(A) = 1/6). So the probability of not rolling a 4 (i.e., the probability of A complement) is 1 – 1/6 = 5/6 (i.e., P(A') = 1 – P(A) = 1 -1/6 = 5/6).

Addition Rules for Probability

Not every event is a simple event; some are compound events (events that can be satisfied in more than one way). For example, if rolling a die, a compound event may be "rolling a 4 or a 5." This event can be satisfied by rolling either a 4 or a 5. The probability of this compound event may be found simply by adding the probability of obtaining a 4 or a 5:

$$P(4 \text{ or } 5) = P(4) + P(5) = \frac{1}{6} + \frac{1}{6} = \frac{2}{6} = \frac{1}{3}.$$

We can generalize this rule by stating P(A or B) = P(A) + P(B). Stay tuned for further information about this formula.

Consider the following experiment: choosing a card from a standard deck of 52 cards. Suppose you would like to find P(Queen or Heart). We can begin this calculation as we did the previous one:

P(Q or H) = P(Q) + P(H) = $\frac{4}{52}+\frac{13}{52}=\frac{17}{52}$. However, if you look at the deck of cards (see above), you might notice that one card – the Queen of Hearts – falls in both categories (Q and H) and thus is double-counted in our probability calculation. We can't have this happen, so to compensate for this double-counting we subtract P(Q and H) = $\frac{1}{52}$; thus,

P(Q or H) = P(Q) + P(H) – P(Q and H) = $\frac{4}{52}+\frac{13}{52}-\frac{1}{52}=\frac{16}{52}=\frac{4}{13}$.

In general, we say that P(A or B) = P(A) + P(B) – P(A and B). This version of the Addition Rule is called the **General Form of the Addition Rule.** It may be used for any "or" probability problem.

Notice we did not subtract anything in the previous example. The reason for this is as follows: There is no one card that can be a 4 and a 5 at the same time. Obtaining a 4 and a 5 on the same single-card draw is impossible; the events "4" and "5" are called ***mutually exclusive*** events because it is impossible for both to happen at the same time; thus, P(4 and 5) =0. When A and B are mutually exclusive, we use the **Special Case of the Addition Rule:** P(A or B) = P(A) + P(B). Consequently, when using the Addition Rule, you first need to decide if events are mutually exclusive, but the General Form will suffice for any "or" situation.

Conditional Probability

Let's look at the standard deck of cards again. The probability of randomly selecting a Queen is P(Q) = $\frac{4}{52}=\frac{1}{13}$. But what if we asked the following: Given we selected a card that is a Heart, what is the probability it is a Queen? The word "given" means that we ***know*** a Heart has been selected (that's the "condition" in "conditional probability"), but we just don't know ***which*** Heart. Recall there are 13 cards in a suit, so the probability we selected a Queen ***given*** we have a Heart is 1/13. In this case, the condition of having selected a Heart *does not affect* the probability of selecting a Queen.

The notation for conditional probability is as follows: P(Q|H). The vertical line is read as the word "given" and the letter following it indicates the condition we know has occurred.

In this example, we saw that P(Q|H) = P(Q) = 1/13, so the condition "Heart" did not affect the probability of getting a Queen.

In this case, we say these events are ***independent*** of each other. In general, when P(A|B) = P(A) we say A and B are ***independent events.*** If P(A|B) ≠ P(A), then the probability of event A has been affected by the condition (event B); thus, A and B are not independent events. They are called ***dependent events.***

The formal definition of conditional probability is as follows: $P(A|B)=\frac{P(A\ and\ B)}{P(B)}$. We can use this definition to compute P(Q|H):

$P(Q|H) = \frac{P(Q \, and \, H)}{P(H)} = \frac{1/52}{13/52} = \frac{1}{52} \div \frac{13}{52} = \frac{1}{52} \cdot \frac{52}{13} = \frac{1}{13}$. This is the same answer we obtained earlier.

Multiplication Rules for Probability

If two events are ***independent*** (as defined above), then $P(A \, and \, B) = P(A) \cdot P(B)$ (this is the Special Multiplication Rule). In other words, simply multiply the probabilities. For example, if we were to roll a die and select a card, the probability of obtaining a "4" and a "Q" is:
$P(4 \, and \, Q) = P(4) \cdot P(Q) = \frac{1}{6} \cdot \frac{1}{13} = \frac{1}{78}$.

If A and B are not independent (i.e., dependent), then we need to take conditional probabilities into account. For example, let's say we have a vase that contains ten marbles: 7 red and 3 blue. If we select two marbles without replacement (i.e., we choose one, don't return it to the vase, and then select the second one) and are interested in *P(Red on first and Red on second)*, the probability of choosing a second red marble is *not* the same as choosing the first red marble because there are fewer red marbles (6) out of the remaining 9 marbles. So,

$$P(Red \, on \, first \, and \, Red \, on \, second) = P(Red \, on \, first) \cdot P(Red \, on \, second | Red \, on \, first)$$
$$= \frac{7}{10} \cdot \frac{6}{9} = \frac{42}{90} = \frac{7}{15}.$$

Thus, the General Multiplication Rule is: $P(A \, and \, B) = P(A) \cdot P(B|A)$
or $P(A \, and \, B) = P(B) \cdot P(A|B)$. Notice that this is simply a rearrangement of the definition of Conditional Probability (see above).

Counting Rules

This may sound odd, but we need to discuss the different counting rules we have in mathematics. These rules will be necessary as we continue our studies in probability.

First, we have the **Fundamental Counting Principle**. In a sequence of n events in which the first event has k_1 possibilities, the second has k_2 possibilities, and so on, and the last one has k_n possibilities, then the total number of possibilities for the sequence of these events can be found by multiplying the number of possibilities at each stage; i.e., $k_1 \cdot k_3 \cdot k_3 \cdots k_n$. This is better understood through an example:

Suppose you were to take a long weekend at the beach. You pack 6 shirts, 4 pair of pants, and 3 jackets. What is the total number of "outfits" you would have with you?

Total number of possibilities = # shirts · # pants · # jackets = 6 · 4 · 3 = 72 "outfits." The reason this works is because for each of the 6 shirts, you have 4 options for pants, giving a total of 24 outfits. For each of these 24 outfits, you have 3 options for jackets, bringing the total to 72 outfits. Simple!

Try this for a three-scoop hot fudge sundae from Somerset Creamery, where they have 33 flavors! Assuming you don't repeat a flavor, you would have 33 · 32 · 31 = 32,736 possibilities! If you could repeat flavors, you would have 33 · 33 · 33 = 33^3 = 35, 937 possibilities!!

Sometimes the order of the possibilities matters; i.e., ABC is different from ACB, BAC, etc. When order matters, we have a *permutation.* There are a couple of ways to compute permutations, but before we do this we need to examine the concept of a *factorial.*

Five factorial, denoted 5!, means that we want to take the product 5! = 5 · 4 · 3 · 2 · 1 = 120. Note that 10! is not twice 5!, because 10! = 10 · 9 · 8 · 7 · 6 · 5 · 4 · 3 · 2 · 1 = 3,628,800.

As an example, let's say we have five photos, but only have room to hang three of them. Further, let's assume the order of the photos matters (perhaps they're photos of your BFFs). How many different orders of these three photos are there?

There are two ways to compute the number of permutations. The first uses the Fundamental Counting Principle: We have five photos from which to choose for the first place; once that one is selected, we have four from which to choose for the second place; then there are three for the third place. Hence, the number of permutations is: $5 \cdot 4 \cdot 3 = 60$ possible ways to hang three photos of the five we have.

Another way is by using the definition of permutation. If we have *n* items and wish to choose *r* of them, we denote this permutation ${}_nP_r$. The formula is as follows:

${}_nP_r = \frac{n!}{(n-r)!}$. In our example of choosing three photos from five, we have n = 5, r = 3, so

${}_5P_3 = \frac{5!}{(5-3)!} = \frac{5!}{2!} = \frac{120}{2} = 60$ possible orderings.

Suppose I ask you to find all possible unique "words" that can be formed using the letters of the word "book." Note that there are two O's here – how would you determine if "book" is different from "book" if we interchange the first and second O's? These would not be "unique" words, so we need to account for that in our calculations. So, we modify the permutation formula to allow for identical objects: the number of possibilities is $\frac{4!}{2!} = \frac{16}{2} = 8$ unique "words." Notice we have 4 letters in the word "book," but we divide out the repetition attributed to the two repeated O's.

If we were to do the same thing for the word "Massachusetts," we note the following: there are 13 letters in Massachusetts; two A's, four S's, and two T's. So, following from the previous example, we have $\frac{13!}{2!4!2!} = 64{,}864{,}800$ unique "words.

If order doesn't matter, then we have *combinations*. With combinations, ABC, ACB, BAC, etc. are all the same combination, so we need to account for that in our calculations. We start with the permutation formula, but we divide out the redundant combinations (much the way we divided out the redundant letters in the two examples above). The combination formula is $_nC_r = \frac{n!}{r!(n-r)!}$.

Suppose we have 5 items from which to choose, and we want to select 3 of them. So, we have $_5C_3 = \frac{5!}{3!(5-3)!} = \frac{5!}{3!\cdot 2!} = \frac{120}{6\cdot 2} = 10$ combinations. The 3! Represents the number of ways we can order three items (ABC, ACB, etc.) that all really represent the same combination. We divide by the 3! so to eliminate redundant orderings.

Powerball

If you've ever played any of the big interstate lottery games, you know the probability of winning is incredibly small! We can calculate the probability of winning by using combinations. Here are the rules: In order to win the jackpot, you need to choose the correct 5 of 69 white balls **and** 1 of 26 red balls. The "and" indicates we need to multiply. In our notation, we have:

$_{69}C_5 \cdot {}_{26}C_1 = \frac{69!}{5!\cdot 64!} \cdot \frac{26!}{1!\cdot 25!} = 11{,}238{,}513 \cdot 26 = 292{,}201{,}338$ combinations. Consequently, should you purchase one ticket, the probability of your winning is $\frac{1}{292{,}201{,}338}$. Not very likely!!

Discrete Probability Distributions

Chapter 5 Summary

The Probability Distribution for a Discrete Random Variable

In statistics, a variable represents some event of interest – the result you get when a die is tossed, etc. There are two types of random variables: continuous and discrete. This chapter is about discrete random variables.

- A discrete random variable has results that come about from a counting process; they are whole numbers (usually). For example, if you were to ask people about the number of siblings they have, their answers would be 0, 1, 2, 3, 4, etc. You cannot have 2.75 siblings!!
- A continuous random variable has results that come from some sort of measurement process and can thus take on fractional (decimal) results. For example, GPAs are not just 0, 1, 2, 3, or 4: you can have a 2.9785 GPA.

A discrete probability distribution consists of the values of the discrete random variable along with the probabilities of each of those values. For example, if we were to roll a fair die, the sample space consists of the values 1, 2, 3, 4, 5, and 6. If we were to combine these values with their respective probabilities (each one is $\frac{1}{6}$ as we have a fair die), we would get the probability distribution:

x	P(x)	xP(x)
1	$\frac{1}{6}$	$\frac{1}{6}$
2	$\frac{1}{6}$	$\frac{2}{6}$
3	$\frac{1}{6}$	$\frac{3}{6}$
4	$\frac{1}{6}$	$\frac{4}{6}$
5	$\frac{1}{6}$	$\frac{5}{6}$
6	$\frac{1}{6}$	$\frac{6}{6}$
	$\sum P(x) = \frac{6}{6} = 1$	$\mu = \sum xP(x) = \frac{21}{6} = 3.5$

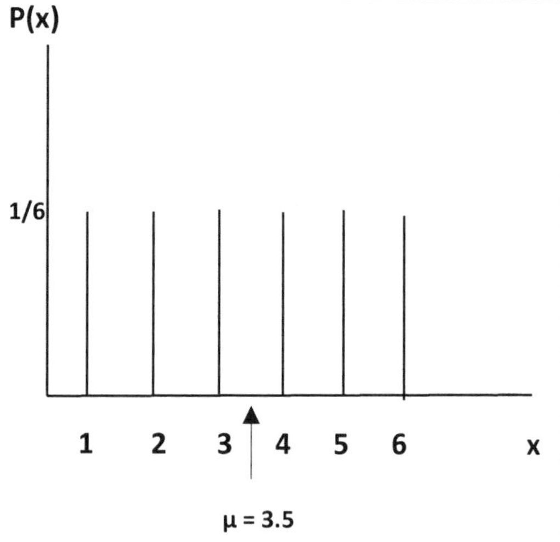

The mean of a discrete probability distribution can be found by taking the sum of the products of each x-value and its probability: $\mu = \sum xP(x)$. The mean of this distribution is 3.5 and has been noted in the graph above. Note that the mean is also called the *expected value, E(x),* since this would be the average outcome you would expect to see after a large (i.e., infinite) number of rolls.

The formulas for variance and standard deviation are similar to the population formulas from Chapter 3:

	Chapter 3 (data)	Chapter 5 (discrete prob. Distr.)
Variance	$\sigma^2 = \frac{\sum(x-\mu)^2}{N}$	$\sigma^2 = \sum[x^2P(x)] - \mu^2$
Standard Deviation	$\sigma = \sqrt{\frac{\sum(x-\mu)^2}{N}}$	$\sigma = \sqrt{\sum[x^2P(x)] - \mu^2}$

So, for our example, $\sigma^2 = \sum[x^2P(x)] - \mu^2 = \frac{91}{6} - (3.5)^2 = \frac{35}{12} \approx 2.91667$; $\sigma = \sqrt{2.91667} \approx 1.7078$.

x	**P(x)**	**x²P(x)**
1	$\frac{1}{6}$	$\frac{1}{6}$
2	$\frac{1}{6}$	$\frac{4}{6}$
3	$\frac{1}{6}$	$\frac{9}{6}$
4	$\frac{1}{6}$	$\frac{16}{6}$
5	$\frac{1}{6}$	$\frac{25}{6}$
6	$\frac{1}{6}$	$\frac{36}{6}$
	$\sum P(x) = \frac{6}{6} = 1$	$\sum x^2P(x) = \frac{91}{6}$

Example: **Raffle Tickets**

The local Boys and Girls club is holding a raffle to raise money for their organization. They will sell 1000 tickets at $1 each and the following prizes will be given: one grand prize of $250; one second prize of $100; and one third prize of $50. If all the tickets are sold and you bought one, what are your expected winnings?

Solution: There are two ways to do this problem. First, let x represent the prize amounts.

x = Amount won	P(x)	xP(x)
250	0.001	0.250
100	0.001	0.100
50	0.001	0.050
0 (you lose!)	0.997	0.000
	1.000	$\mu = \sum xP(x) = 0.400$

Note that μ = \$0.40. This is logical, since the organization will give away a total of \$400 among the 1000 ticket holders, so the average winnings per person is $\frac{\$400}{1000} = \$0.40$. To calculate your expected winnings, take the average winnings of \$0.40 and subtract the cost of your ticket (\$1). Your expected winnings are \$-0.60. (Again, this is logical since the organization doesn't give out \$600, so the average person loses \$0.60.)

Another way to do this problem is to let x represent the *net* winnings; that is, the winnings after the cost of the ticket is taken into consideration. Thus, μ represents the expected *net* winnings: μ = \$-0.60.

x = net won	P(x)	xP(x)
249	0.001	0.249
99	0.001	0.099
49	0.001	0.049
-1 (you lose!)	0.997	-0.997
	1.000	$\mu = \sum xP(x) = -0.60$

Special Probability Distributions

There are several discrete probability distributions that help explain probabilities in "real-world" situations. We will examine two of these: the Binomial and Poisson Distributions.

Binomial Probability Distribution

As the name implies, the binomial distribution applies to situations in which there are only two possible outcomes. These are traditionally called "success" and "failure" (although some texts refer to these as the "event of interest" and "not the event of interest).

In order to use the binomial distribution, you must be certain you have ***Bernoulli Trials.*** Bernoulli Trials have the following characteristics:

- There are n identical, independent trials with only two outcomes: "success" or "failure"
- The probability of success and failure are the same on each trial
- P(Success) = p and P(Failure) = q; note: p + q = 1
- We want x successes (and hence (n – x) failures)

The binomial probability formula for P(x successes) is: $P(x\ succsses) = \ _nC_x \cdot p^x \cdot q^{n-x}$.

The combination coefficient, $_nC_x$, tells us the number of ways we can have x successes; p^x gives the probability of getting x successes; q^{n-x} gives the probability of getting (n – x) failures.

Example: **Voting**

Assume that 60% of college students tend to vote for political party A and 40% vote for party B. If 5 students are selected at random, find the probability that exactly 3 of them will vote for party A.

Solution:
Here we have p = 0.60, q = 0.40, n = 5, and x = 3. Note that if we have 3 successes, we must have 2 failures.

So $P(x = 3\ "succsses") = \ _5C_3 \cdot (.6)^3 \cdot (.4)^2$
$= (10)(0.216)(0.16) = 0.3456.$

We can also find this probability by looking at a binomial probability table where n = 5, x = 3 and p = 0.60. These tables can easily be found online. (Note: this table uses *r* rather than *x* to denote number of successes.) Beware: some Binomial tables contain cumulative probabilities in each column, not the individual probabilities.

n = 5

p	.01	.05	.10	.15	.20	.25	.30	.35	.40	.45	.50	.55	.60	.65	.70	.75	.80	.85	.90	.95
r 0	.951	.774	.590	.444	.328	.237	.168	.116	.078	.050	.031	.019	.010	.005	.002	.001	.000	.000	.000	.000
1	.048	.204	.328	.392	.410	.396	.360	.312	.259	.206	.156	.113	.077	.049	.028	.015	.006	.002	.000	.000
2	.001	.021	.073	.138	.205	.264	.309	.336	.346	.337	.312	.276	.230	.181	.132	.088	.051	.024	.008	.001
3	.000	.001	.008	.024	.051	.088	.132	.181	.230	.276	.312	.337	.346	.336	.309	.264	.205	.138	.073	.021
4	.000	.000	.000	.002	.006	.015	.028	.049	.077	.113	.156	.206	.259	.312	.360	.396	.410	.392	.328	.204
5	.000	.000	.000	.000	.000	.001	.002	.005	.010	.019	.031	.050	.078	.116	.168	.237	.328	.444	.590	.774

A third way of finding this probability is by using your TI graphing calculator. The step-by-step directions can be found in this module in eLearning. Choose 2nd → VARS to reach the Distr(ibution) menu; Choose option A: Binompdf (binomial probability density function). The syntax is as follows: Binompdf(n, p, x); in this case, Binompdf(5, 0.60, 3). After hitting ENTER, you should see the result we calculated.

Example, con't.

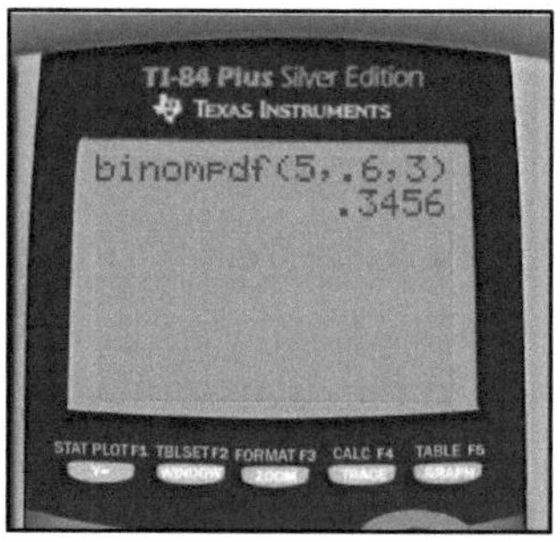

Example:

Now suppose we are interested in finding *at most 3* students who will vote for Party A.

Solution:

In this case, we are interested in P(x $\leq$ 3). This means we want P(x = 0 or x = 1 or x = 2 or x = 3), a total of four calculations. This is tedious, but doable; however, there are two ways to avoid doing direct calculations. A second way is by referring to the table above. In this case, we want the sum of the first four probabilities in the 0.60 column: 0.010 + 0.077 + 0.230 + 0.346 = 0.663.

A third way to find this probability is by using your TI calculator. In this case, choose option B:Binomcdf (binomial cumulative density function – this function gives the sum of the probabilities from x = 0 to x = the value entered). Binomcdf(5, 0.60, 3). After hitting ENTER, you should see the result we calculated.

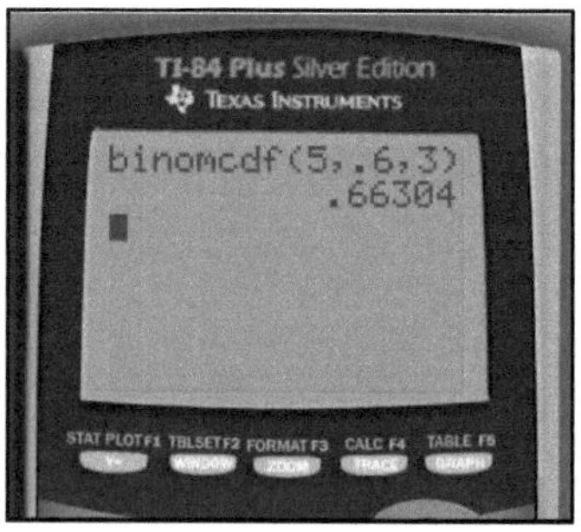

Now suppose we are interested in finding *more than* 3 students who will vote for Party A.

Solution:

P(x > 3) is the complement of P(x $\leq$ 3) which we found above. Hence, we can use the fact that the sum of complementary events is 1; thus P(x > 3) = 1 - P(x $\leq$ 3) = 1 – 0.663 = 0.337.

Elaine A. Previte
May 2019

Poisson Probability Distribution

Another useful discrete probability distribution is the *Poisson* distribution. The Poisson gives the probability of a number of events occurring, but usually they occur per unit of time or some other measure. Poisson requires the average number of occurrences per unit of time; however, unlike the Binomial distribution, there is no upper limit on the number of occurrences that may occur per unit of time. For example, if a fast-food outlet *averages* 10 cars at the drive-through window per 15-minute period during lunchtime, then the number of cars that actually show up can well exceed 10. In theory, there is no upper limit on the number of occurrences, although the probabilities become quite small for the higher numbers.

The formula for the Poisson distribution is as follows: $P(x|\lambda) = \frac{e^{-\lambda}\lambda^x}{x!}$, where x is the number of occurrences, λ is the average number of occurrences per unit of time (or other measure), and e is a constant (e ≈ 2.718271828459...). It is best to use a table or a TI to compute Poisson probabilities.

Example: Drive-thru Window

Let's say the local fast-food restaurant, *Mc JunkFood,* averages 10 cars every 15 minutes during the lunchtime rush. Assuming it is lunchtime, find the probability that exactly 10 cars arrive at the window within the next 15 minutes.

Solution:

Here $\lambda = 10$ and x = 10, so we are looking for $P(x = 10|\ \lambda = 10)$. Simply look for the intersection of the $\lambda = 10$ column and the x = 10 row. $P(x = 10|\ \lambda = 10) = 0.125$. (Note: empty cells indicate that the probability for that cell would be zero when rounded to four decimal places.)

k \ λ	0,2	0,3	0,4	0,5	0,6	1	1,5	2	3	4	5	6	7	8	9	10
0	0,8187	0,7408	0,6703	0,6065	0,5488	0,368	0,223	0,135	0,050	0,018	0,007	0,002	0,001	0,000	0,000	0,000
1	0,1637	0,2222	0,2681	0,3033	0,3293	0,368	0,335	0,271	0,149	0,073	0,034	0,015	0,006	0,003	0,001	0,000
2	0,0164	0,0333	0,0536	0,0758	0,0988	0,184	0,251	0,271	0,224	0,147	0,084	0,045	0,022	0,011	0,005	0,002
3	0,0011	0,0033	0,0072	0,0126	0,0198	0,061	0,126	0,180	0,224	0,195	0,140	0,089	0,052	0,029	0,015	0,008
4	0,0001	0,0003	0,0007	0,0016	0,0030	0,015	0,047	0,090	0,168	0,195	0,175	0,134	0,091	0,057	0,034	0,019
5		0,0000	0,0001	0,0002	0,0004	0,003	0,014	0,036	0,101	0,156	0,175	0,161	0,128	0,092	0,061	0,038
6			0,0000	0,0000	0,0000	0,001	0,004	0,012	0,050	0,104	0,146	0,161	0,149	0,122	0,091	0,063
7						0,000	0,001	0,003	0,022	0,060	0,104	0,138	0,149	0,140	0,117	0,090
8							0,000	0,001	0,008	0,030	0,065	0,103	0,130	0,140	0,132	0,113
9								0,000	0,003	0,013	0,036	0,069	0,101	0,124	0,132	~~0,125~~
10									0,001	0,005	0,018	0,041	0,071	0,099	0,119	0,125
11									0,000	0,002	0,008	0,023	0,045	0,072	0,097	~~0,114~~
12										0,001	0,003	0,011	0,026	0,048	0,073	0,095
13										0,000	0,001	0,005	0,014	0,030	0,050	0,073
14											0,000	0,002	0,007	0,017	0,032	0,052
15												0,001	0,003	0,009	0,019	0,035
16												0,000	0,001	0,005	0,011	0,022
17													0,001	0,002	0,006	0,013
18													0,000	0,001	0,003	0,007
19														0,000	0,001	0,004
20															0,001	0,002
21															0,000	0,001
22																0,000

Notice that this table continues well past 10 occurrences… the probability of 21 occurrences is 0.001 (rounded to four places), so it is possible (although not probable) that 21 cars could arrive at the drive-thru during a 15-minute period during lunchtime!

Note: if you were interested in the number of occurrences during, say, 30 minutes, you would need to double the value of λ to 20 and use that column in a Poisson table.

When using the TI, again go to the Distributions menu (2nd → VARS) and choose option C:Poissonpdf. The syntax is Poissonpdf(λ, x); in this case, Poissonpdf(10, 10). You should see the results below:

Suppose we are interested in the probability that 10 or fewer cars arrive at the window during a 15-minute period. This would be P(x ≤ 10), so we could either add the probabilities for x = 0, 1, 2, …, 10 or use a cumulative Poisson table. We could also use option D:Poissoncdf under the DISTR menu on our TIs. We would see the following results:

Should we want P(x > 10), we would use the fact that P(x>10) is the complement of P(x≤10):
P(x > 10) = 1 – P(x ≤ 10) = 1 – 0.5830 = 0.4170.

Normal Probability Distributions
Chapter 6 Summary

In Chapter 5, we discussed discrete probability distributions; that is, probability distributions that were based on discrete random variables. Now we will look at some probability distributions that are based on continuous random variables. While there are several continuous probability distributions, we will focus primarily on the normal distribution.

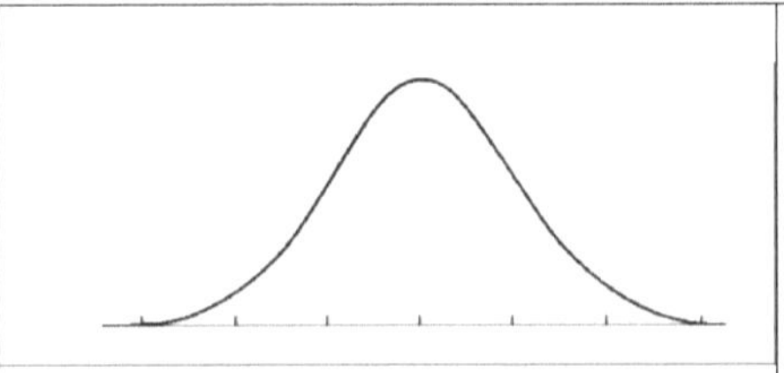

Normal Distribution

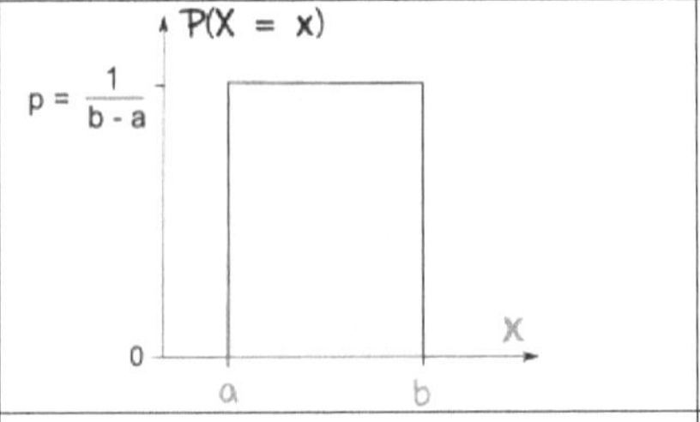

Uniform Distribution

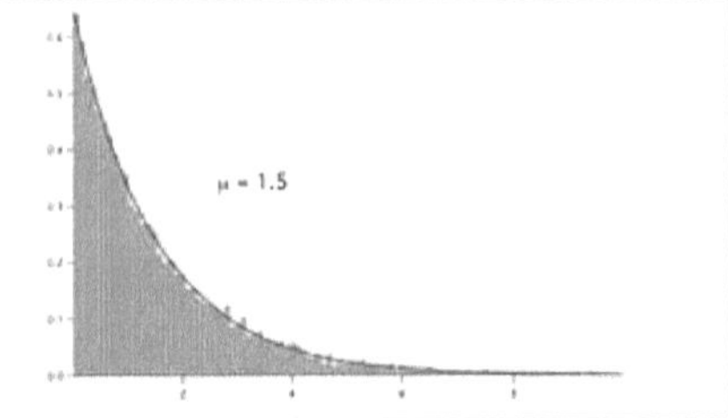

Exponential Distribution

Many phenomena in the real world end up following a normal distribution; hence it is the most important distribution for our purposes. Characteristics of the distribution are:

- It is mounded (bell-shaped)
- Total area under the curve is 1
- It is symmetrical about the mean; therefore, its mean, median and mode are all equal and 50% of the area is below the mean; 50% above
- Its IQR is 1.33 $\left(i.e., \frac{4}{3}\right)$standard deviations; therefore, the middle 50% of the distribution lies within $\mu \pm \frac{2}{3}\sigma$
- Its domain is (theoretically) infinite (i.e., $-\infty < x < \infty$)

Since the normal distribution is so important and so widely used, statisticians have developed several ways to help us find the probabilities we need. Consequently, you will never have to deal with the actual mathematics of the function. Be grateful, because the probability density function (PDF) that defines the normal curve is:

$$y = P(x) = \frac{e^{\frac{-(x-\mu)^2}{2\sigma^2}}}{\sigma\sqrt{2\pi}}.$$

This function was not put here to frighten you! Rather, it is here to help illustrate a couple of important points:

- As expected, x is the independent variable; y is the dependent variable
- π is the constant from geometry that you're accustomed to: π ≈ 3.141592654....
- e is the base of the natural logarithm: e ≈ 2.718281828459...

More importantly, notice the following:

- the population mean, μ, is part of the function
- the population standard deviation, σ, is also included.

While this may mean nothing to you, you should note that the last two points imply that, for every different set of values of μ and σ, you get a slightly different version of the normal PDF. This means there are ***an infinite number*** of normal PDFs. And, if you ever took a basic business calculus course, you know that we find the area under the curve (i.e., the probability of a < x < b) as the definite integral of the PDF from a to b, so you would have to integrate the PDF *every time* you wanted to find probabilities. If this goes over your mathematical head, worry not! Read on!

To help simplify things, statisticians use the standard normal curve by setting μ = 0 and σ = 1. Notice how that helps to "clean up" the normal PDF:

$$y = P(x) = \frac{e^{\frac{-z^2}{2}}}{\sqrt{2\pi}}.$$

Now, this is still an intimidating function, but it has helped statisticians develop standard normal tables to assist in finding probabilities... Much easier than integrating this function!

So, now let's go back to the uses of the normal distribution and why the standard normal PDF is so useful.

Many things follow a normal distribution:

- heights and weights of people
- Blood pressure
- IQs
- Test scores
- Dress sizes
- Size of widgets produced by machines
- Etc.

For example, a particular IQ test has a mean of 100 with a standard deviation of 15. Suppose you have an IQ of 115: where are you on the curve?

You can see that an IQ of 115 lies to the right of the mean and that it is one standard deviation above the mean. You can standardize the IQ score by using the z-score formula (see chapter 3): $$z = \frac{x-\mu}{\sigma} = \frac{115-100}{15} = \frac{15}{15} = 1$$ All IQ scores can be standardized, and the resulting standard scores are included below the raw IQ scores.	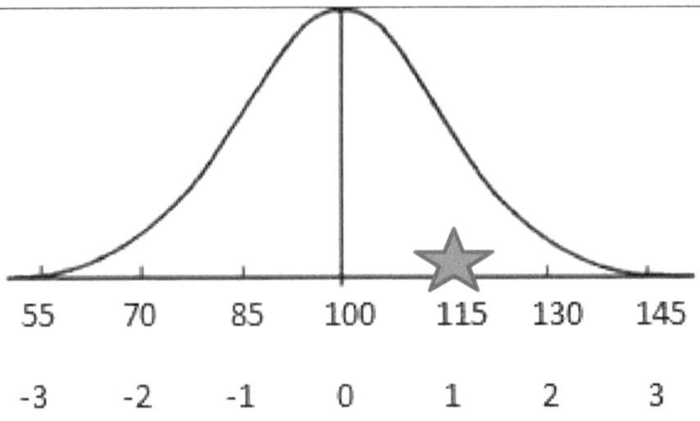

As another example, suppose a soda-filling machine fills bottles with a mean of 20 fl. oz. and a standard deviation of 0.10 fl. oz. What is the z-score of a bottle that contains 19.8 fl. oz.?

A volume of 19.8 fl. oz. is below the mean of 20 fl. oz.; thus, its z-score must be negative: $$z = \frac{x-\mu}{\sigma} = \frac{19.8-20}{0.10} = \frac{-0.2}{0.1} = -2$$ All volumes of soda bottles can be standardized, and the resulting standard scores are included below the raw soda volumes.	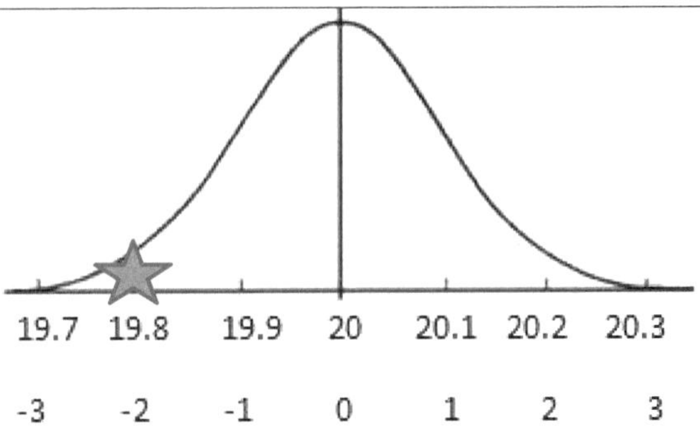

Area under the Curve = Probability!!

Now, suppose you're interested in the probability of finding a person whose IQ is less than 115. From the z-score formula, you can see that an IQ of 115 equates to a z-score of 1. Again, from Calculus, to find the probability of finding a person in this range of IQ scores, you would have to integrate the Normal PDF from $-\infty$ to 115, or the Standard Normal PDF from $-\infty$ to1. Not going to happen!

One way to find the probability is to use a Standard Normal Table, a portion of which is pictured below.

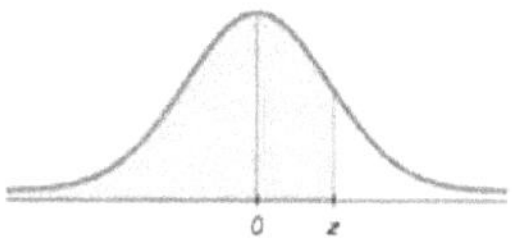

POSITIVE z Scores

TABLE A-2 (*continued*) Cumulative Area from the LEFT

z	.00	.01	.02	.03	.04	.05	.06	.07	.08	.09
0.0	.5000	.5040	.5080	.5120	.5160	.5199	.5239	.5279	.5319	.5359
0.1	.5398	.5438	.5478	.5517	.5557	.5596	.5636	.5675	.5714	.5753
0.2	.5793	.5832	.5871	.5910	.5948	.5987	.6026	.6064	.6103	.6141
0.3	.6179	.6217	.6255	.6293	.6331	.6368	.6406	.6443	.6480	.6517
0.4	.6554	.6591	.6628	.6664	.6700	.6736	.6772	.6808	.6844	.6879
0.5	.6915	.6950	.6985	.7019	.7054	.7088	.7123	.7157	.7190	.7224
0.6	.7257	.7291	.7324	.7357	.7389	.7422	.7454	.7486	.7517	.7549
0.7	.7580	.7611	.7642	.7673	.7704	.7734	.7764	.7794	.7823	.7852
0.8	.7881	.7910	.7939	.7967	.7995	.8023	.8051	.8078	.8106	.8133
0.9	.8159	.8186	.8212	.8238	.8264	.8289	.8315	.8340	.8365	.8389
1.0	.8413	.8438	.8461	.8485	.8508	.8531	.8554	.8577	.8599	.8621
1.1	.8643	.8665	.8686	.8708	.8729	.8749	.8770	.8790	.8810	.8830
1.2	.8849	.8869	.8888	.8907	.8925	.8944	.8962	.8980	.8997	.9015

This page gives the **cumulative** area under the curve (i.e., the area from $-\infty$ to the z-score you're looking up).

The integer and first decimal place of the z-score are located in the left-hand column and the second places can be found be looking at the column headings. So the cumulative area to a z-score of 1.00 can be found by looking at the intersection of the 1.0 row and the 0.00 column.

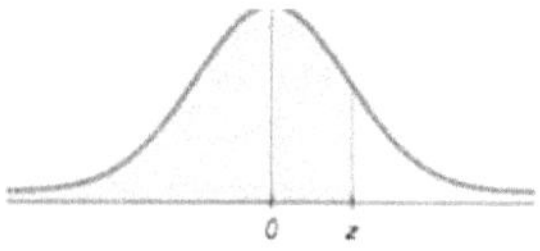

POSITIVE z Scores

TABLE A-2 (*continued*) Cumulative Area from the LEFT

z	.00	.01	.02	.03	.04	.05	.06	.07	.08	.09
0.0	.5000	.5040	.5080	.5120	.5160	.5199	.5239	.5279	.5319	.5359
0.1	.5398	.5438	.5478	.5517	.5557	.5596	.5636	.5675	.5714	.5753
0.2	.5793	.5832	.5871	.5910	.5948	.5987	.6026	.6064	.6103	.6141
0.3	.6179	.6217	.6255	.6293	.6331	.6368	.6406	.6443	.6480	.6517
0.4	.6554	.6591	.6628	.6664	.6700	.6736	.6772	.6808	.6844	.6879
0.5	.6915	.6950	.6985	.7019	.7054	.7088	.7123	.7157	.7190	.7224
0.6	.7257	.7291	.7324	.7357	.7389	.7422	.7454	.7486	.7517	.7549
0.7	.7580	.7611	.7642	.7673	.7704	.7734	.7764	.7794	.7823	.7852
0.8	.7881	.7910	.7939	.7967	.7995	.8023	.8051	.8078	.8106	.8133
0.9	.8159	.8186	.8212	.8238	.8264	.8289	.8315	.8340	.8365	.8389
1.0	.8413	.8438	.8461	.8485	.8508	.8531	.8554	.8577	.8599	.8621
1.1	.8643	.8665	.8686	.8708	.8729	.8749	.8770	.8790	.8810	.8830
1.2	.8849	.8869	.8888	.8907	.8925	.8944	.8962	.8980	.8997	.9015

The circled value, 0.8413, gives the cumulative area under the normal curve from $-\infty$ to 1. This means that there is an 84.13% probability of encountering someone with an IQ of 115 or less. Additionally, you can interpret this to mean that 84.13% of the population has IQs of 115 or less.

Now, suppose you're interested in the probability of finding a person whose IQ lies between 85 and 115. From the z-score formula, you can see that an IQ of 85 equates to a z-score of -1, and a 115 equates to a +1. Using the Standard Normal Table below, we can see that a z-score of -1.00 is on the left-hand page, and has an entry of 0.1587. This means that there is 15.87% of the area under the curve between $-\infty$ and -1. Using the positive z-score page, we see again that there is an entry of 0.8413 at z = +1. Again, this means that 84.13% of the area under the curve is between $-\infty$ and +1. Since we want the area between -1 and +1, we subtract the smaller area from the larger area. The result will be the area between -1 and +1.

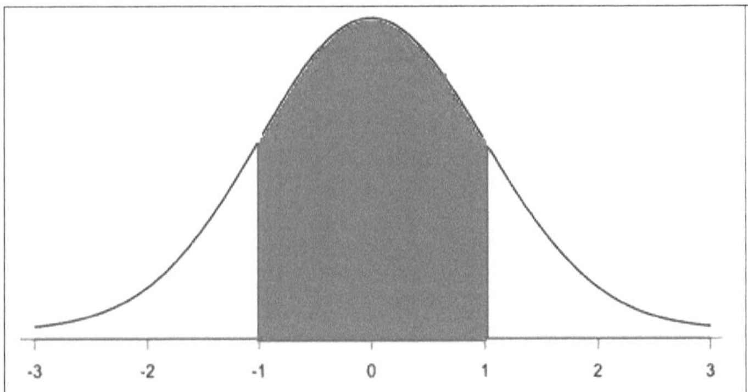

$P(85 < x < 115) = P(-1 < z < 1)$
$= 0.8413 - 0.1587$
$= 0.6826$

Recall that the Empirical Rule stated that approximately 68% of the area under the normal curve lies between -1 and +1. Can you now see how they arrived at that value?

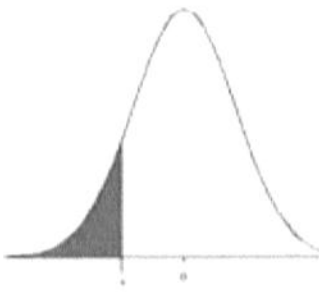

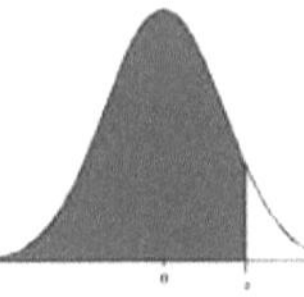

z	0.00	0.01	0.02	0.03	0.04	0.05	0.06	0.07	0.08	0.09
-3.4	0.0003	0.0003	0.0003	0.0003	0.0003	0.0003	0.0003	0.0003	0.0003	0.0002
-3.3	0.0005	0.0005	0.0005	0.0004	0.0004	0.0004	0.0004	0.0004	0.0004	0.0003
-3.2	0.0007	0.0007	0.0006	0.0006	0.0006	0.0006	0.0006	0.0005	0.0005	0.0005
-3.1	0.0010	0.0009	0.0009	0.0009	0.0008	0.0008	0.0008	0.0008	0.0007	0.0007
-3.0	0.0013	0.0013	0.0013	0.0012	0.0012	0.0011	0.0011	0.0011	0.0010	0.0010
-2.9	0.0019	0.0018	0.0018	0.0017	0.0016	0.0016	0.0015	0.0015	0.0014	0.0014
-2.8	0.0026	0.0025	0.0024	0.0023	0.0023	0.0022	0.0021	0.0021	0.0020	0.0019
-2.7	0.0035	0.0034	0.0033	0.0032	0.0031	0.0030	0.0029	0.0028	0.0027	0.0026
-2.6	0.0047	0.0045	0.0044	0.0043	0.0041	0.0040	0.0039	0.0038	0.0037	0.0036
-2.5	0.0062	0.0060	0.0059	0.0057	0.0055	0.0054	0.0052	0.0051	0.0049	0.0048
-2.4	0.0082	0.0080	0.0078	0.0075	0.0073	0.0071	0.0069	0.0068	0.0066	0.0064
-2.3	0.0107	0.0104	0.0102	0.0099	0.0096	0.0094	0.0091	0.0089	0.0087	0.0084
-2.2	0.0139	0.0136	0.0132	0.0129	0.0125	0.0122	0.0119	0.0116	0.0113	0.0110
-2.1	0.0179	0.0174	0.0170	0.0166	0.0162	0.0158	0.0154	0.0150	0.0146	0.0143
-2.0	0.0228	0.0222	0.0217	0.0212	0.0207	0.0202	0.0197	0.0192	0.0188	0.0183
-1.9	0.0287	0.0281	0.0274	0.0268	0.0262	0.0256	0.0250	0.0244	0.0239	0.0233
-1.8	0.0359	0.0351	0.0344	0.0336	0.0329	0.0322	0.0314	0.0307	0.0301	0.0294
-1.7	0.0446	0.0436	0.0427	0.0418	0.0409	0.0401	0.0392	0.0384	0.0375	0.0367
-1.6	0.0548	0.0537	0.0526	0.0516	0.0505	0.0495	0.0485	0.0475	0.0465	0.0455
-1.5	0.0668	0.0655	0.0643	0.0630	0.0618	0.0606	0.0594	0.0582	0.0571	0.0559
-1.4	0.0808	0.0793	0.0778	0.0764	0.0749	0.0735	0.0721	0.0708	0.0694	0.0681
-1.3	0.0968	0.0951	0.0934	0.0918	0.0901	0.0885	0.0869	0.0853	0.0838	0.0823
-1.2	0.1151	0.1131	0.1112	0.1093	0.1075	0.1056	0.1038	0.1020	0.1003	0.0985
-1.1	0.1357	0.1335	0.1314	0.1292	0.1271	0.1251	0.1230	0.1210	0.1190	0.1170
-1.0	0.1587	0.1562	0.1539	0.1515	0.1492	0.1469	0.1446	0.1423	0.1401	0.1379
-0.9	0.1841	0.1814	0.1788	0.1762	0.1736	0.1711	0.1685	0.1660	0.1635	0.1611
-0.8	0.2119	0.2090	0.2061	0.2033	0.2005	0.1977	0.1949	0.1922	0.1894	0.1867
-0.7	0.2420	0.2389	0.2358	0.2327	0.2296	0.2266	0.2236	0.2206	0.2177	0.2148
-0.6	0.2743	0.2709	0.2676	0.2643	0.2611	0.2578	0.2546	0.2514	0.2483	0.2451
-0.5	0.3085	0.3050	0.3015	0.2981	0.2946	0.2912	0.2877	0.2843	0.2810	0.2776
-0.4	0.3446	0.3409	0.3372	0.3336	0.3300	0.3264	0.3228	0.3192	0.3156	0.3121
-0.3	0.3821	0.3783	0.3745	0.3707	0.3669	0.3632	0.3594	0.3557	0.3520	0.3483
-0.2	0.4207	0.4168	0.4129	0.4090	0.4052	0.4013	0.3974	0.3936	0.3897	0.3859
-0.1	0.4602	0.4562	0.4522	0.4483	0.4443	0.4404	0.4364	0.4325	0.4286	0.4247
-0.0	0.5000	0.4960	0.4920	0.4880	0.4840	0.4801	0.4761	0.4721	0.4681	0.4641

z	0.00	0.01	0.02	0.03	0.04	0.05	0.06	0.07	0.08	0.09
0.0	0.5000	0.5040	0.5080	0.5120	0.5160	0.5199	0.5239	0.5279	0.5319	0.5359
0.1	0.5398	0.5438	0.5478	0.5517	0.5557	0.5596	0.5636	0.5675	0.5714	0.5753
0.2	0.5793	0.5832	0.5871	0.5910	0.5948	0.5987	0.6026	0.6064	0.6103	0.6141
0.3	0.6179	0.6217	0.6255	0.6293	0.6331	0.6368	0.6406	0.6443	0.6480	0.6517
0.4	0.6554	0.6591	0.6628	0.6664	0.6700	0.6736	0.6772	0.6808	0.6844	0.6879
0.5	0.6915	0.6950	0.6985	0.7019	0.7054	0.7088	0.7123	0.7157	0.7190	0.7224
0.6	0.7257	0.7291	0.7324	0.7357	0.7389	0.7422	0.7454	0.7486	0.7517	0.7549
0.7	0.7580	0.7611	0.7642	0.7673	0.7704	0.7734	0.7764	0.7794	0.7823	0.7852
0.8	0.7881	0.7910	0.7939	0.7967	0.7995	0.8023	0.8051	0.8078	0.8106	0.8133
0.9	0.8159	0.8186	0.8212	0.8238	0.8264	0.8289	0.8315	0.8340	0.8365	0.8389
1.0	0.8413	0.8438	0.8461	0.8485	0.8508	0.8531	0.8554	0.8577	0.8599	0.8621
1.1	0.8643	0.8665	0.8686	0.8708	0.8729	0.8749	0.8770	0.8790	0.8810	0.8830
1.2	0.8849	0.8869	0.8888	0.8907	0.8925	0.8944	0.8962	0.8980	0.8997	0.9015
1.3	0.9032	0.9049	0.9066	0.9082	0.9099	0.9115	0.9131	0.9147	0.9162	0.9177
1.4	0.9192	0.9207	0.9222	0.9236	0.9251	0.9265	0.9279	0.9292	0.9306	0.9319
1.5	0.9332	0.9345	0.9357	0.9370	0.9382	0.9394	0.9406	0.9418	0.9429	0.9441
1.6	0.9452	0.9463	0.9474	0.9484	0.9495	0.9505	0.9515	0.9525	0.9535	0.9545
1.7	0.9554	0.9564	0.9573	0.9582	0.9591	0.9599	0.9608	0.9616	0.9625	0.9633
1.8	0.9641	0.9649	0.9656	0.9664	0.9671	0.9678	0.9686	0.9693	0.9699	0.9706
1.9	0.9713	0.9719	0.9726	0.9732	0.9738	0.9744	0.9750	0.9756	0.9761	0.9767
2.0	0.9772	0.9778	0.9783	0.9788	0.9793	0.9798	0.9803	0.9808	0.9812	0.9817
2.1	0.9821	0.9826	0.9830	0.9834	0.9838	0.9842	0.9846	0.9850	0.9854	0.9857
2.2	0.9861	0.9864	0.9868	0.9871	0.9875	0.9878	0.9881	0.9884	0.9887	0.9890
2.3	0.9893	0.9896	0.9898	0.9901	0.9904	0.9906	0.9909	0.9911	0.9913	0.9916
2.4	0.9918	0.9920	0.9922	0.9925	0.9927	0.9929	0.9931	0.9932	0.9934	0.9936
2.5	0.9938	0.9940	0.9941	0.9943	0.9945	0.9946	0.9948	0.9949	0.9951	0.9952
2.6	0.9953	0.9955	0.9956	0.9957	0.9959	0.9960	0.9961	0.9962	0.9963	0.9964
2.7	0.9965	0.9966	0.9967	0.9968	0.9969	0.9970	0.9971	0.9972	0.9973	0.9974
2.8	0.9974	0.9975	0.9976	0.9977	0.9977	0.9978	0.9979	0.9979	0.9980	0.9981
2.9	0.9981	0.9982	0.9982	0.9983	0.9984	0.9984	0.9985	0.9985	0.9986	0.9986
3.0	0.9987	0.9987	0.9987	0.9988	0.9988	0.9989	0.9989	0.9989	0.9990	0.9990
3.1	0.9990	0.9991	0.9991	0.9991	0.9992	0.9992	0.9992	0.9992	0.9993	0.9993
3.2	0.9993	0.9993	0.9994	0.9994	0.9994	0.9994	0.9994	0.9995	0.9995	0.9995
3.3	0.9995	0.9995	0.9995	0.9996	0.9996	0.9996	0.9996	0.9996	0.9996	0.9997
3.4	0.9997	0.9997	0.9997	0.9997	0.9997	0.9997	0.9997	0.9997	0.9997	0.9998

Suppose you want to know the probability of finding a person whose IQ is greater than 115. That means you want his/her z-score to be greater than +1, and thus want the area under the curve that lies between +1 and ∞.

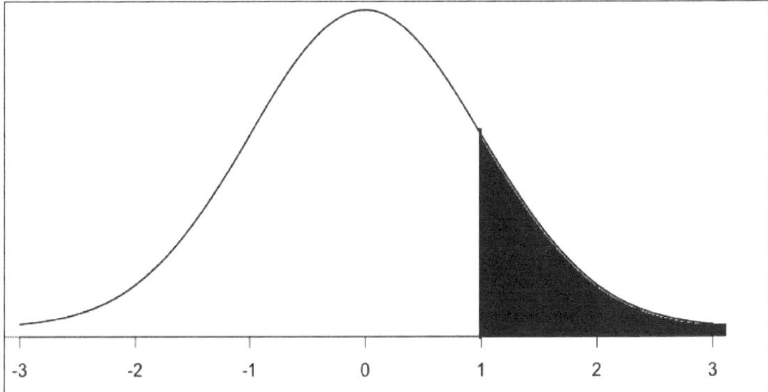

Since the table gives cumulative area from the left and you wish to go to the right, you need the complement of the value given for $z = +1.00$, so

$P(z > 1) = 1 - P(z \leq 1)$
$= 1 - 0.8413$
$= 0.1587.$

This means that 15.87% of the population has IQs greater than 115.

Finding x

Now let's look at the Jeopardy version of a normal curve problem; that is, you know something about the area you want, but you don't know which value of x gets you there.

For example, say a particular statistics class' final grades are normally distributed and have a mean score of 80 with a standard deviation of 9. You're hoping to be in the top 20% of the class. What would be the final average you need to have to be there?

First, you need to understand the information given to you... The mean of 80 tells us that $\mu = 80$. A standard deviation of 9 means that $\sigma = 9$. So far, so good!

Now, understand the remaining information. You wish to be in the top 20% of the class. Imagine where that would put you with respect to the normal curve. Top 20% → Highest 20% of the area under the normal curve, correct? This location and z-score is also the cut-off for the lowest 80% of students.

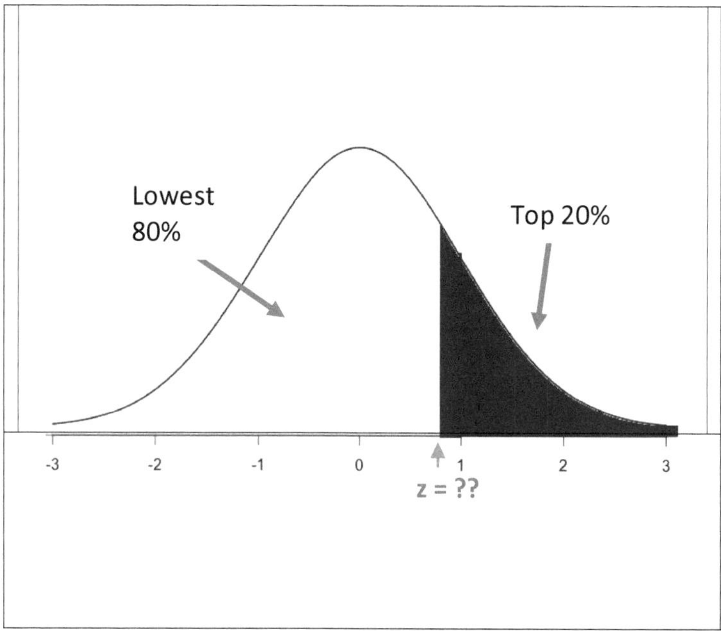

Since the tables give cumulative area from the left, you need to look for the area closest to 0.80. The z-score that corresponds to the lowest 80% of the area is z = 0.84.

Using the z-score formula we have:

$$z = \frac{x - \mu}{\sigma}$$

$$0.84 = \frac{x - 80}{9}.$$

Solving for x, we get:

$$x = 9(0.84) + 80 = 87.56.$$

This means that the top 20% of students have grades of 87.56 or above.

In general, when z, μ and σ are known, you may use the formula $\boldsymbol{x = z\sigma + \mu}$ to solve for x.

Sampling Distributions
Chapter 7 Summary

In business as well as other areas, one does not simply collect data for data's sake: there has to be a purpose for expending the resources necessary to do data collection and analysis. And there is a reason: businesses want to use currently available data to make inferences about what's happening in the market and then to make projections that will inform their decisions for the near future.

We'll begin by talking about the ***sampling distribution of the sample means***.

When you collect a sample of data, the most common descriptive measure you calculate is the sample mean, $\overline{x}$. But that particular $\overline{x}$ is simply one of many possible sample means you could have found when taking samples of the same size from the population under consideration.

By way of illustration, let's assume your evil statistics instructor has a series of four quizzes for each chapter in the course. Each quiz A has 1 question; B has 2; C has 3; and D has 4. If you have to take quizzes for the next two chapters, how many questions could you get? In other words, what are the possible combination of quizzes you can get? And what would be the corresponding numbers of questions you would answer?

First, note that the population is a uniform discrete distribution; i.e., each quiz has a 0.25 probability of being selected for any given chapter. The mean, or expected value, of this distribution is 2.5 questions, and the standard deviation is 1.118. (Recall from Chapter 5 that the formula for the mean of a discrete probability distribution is $\mu = E(x) = \sum xP(x)$ and the formula for the standard deviation is

$$\sigma = \sqrt{\Sigma\left(x^2P(x)\right) - \mu^2}\ ).$$

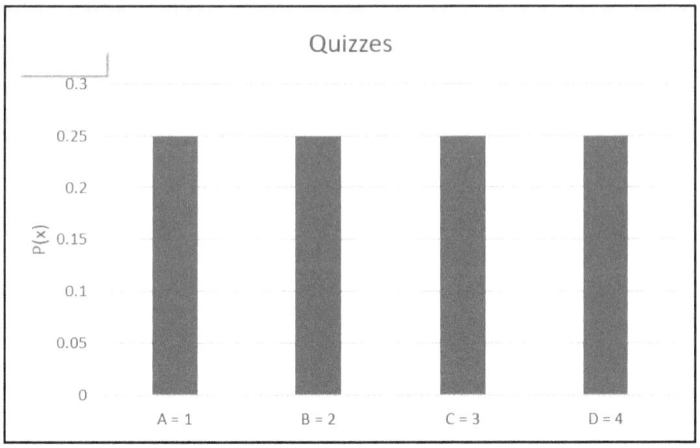

Since you're taking two quizzes, we're looking at all possible samples of size 2 from the population of N = 4 possible quizzes.

Elaine A. Previte
May 2017

Samples of size n = 2 i.e., Quiz 1, Quiz 2	Numbers of Questions Per Quiz	Sample Means
A, A	1, 1	1
A, B	1, 2	1.5
A, C	1, 3	2
A, D	1, 4	2.5
B, A	2, 1	1.5
B, B	2, 2	2
B, C	2, 3	2.5
B, D	2, 4	3
C, A	3, 1	2
C, B	3, 2	2.5
C, C	3, 3	3
C, D	3, 4	3.5
D, A	4, 1	2.5
D, B	4, 2	3
D, C	4, 3	3.5
D, D	4, 4	4

If we look at all possible sample means, along with their relative frequencies (i.e., the probability each one may be selected), we will get the ***sampling distribution of sample means***. Notice the graph of this sampling distribution: it is a mounded (i.e., bell-shaped or normal) distribution, with a mean of 2.25 and a standard deviation of 0.7906. Do you find this surprising, considering the population was uniformly distributed (i.e., "flat")?

$\bar{x}$	$P(\bar{x})$
1.0	$\frac{1}{16} = 0.0625$
1.5	$\frac{2}{16} = 0.1250$
2.0	$\frac{3}{16} = 0.1875$
2.5	$\frac{4}{16} = 0.2500$
3.0	$\frac{3}{16} = 0.1875$
3.5	$\frac{2}{16} = 0.1250$
4.0	$\frac{1}{16} = 0.0625$

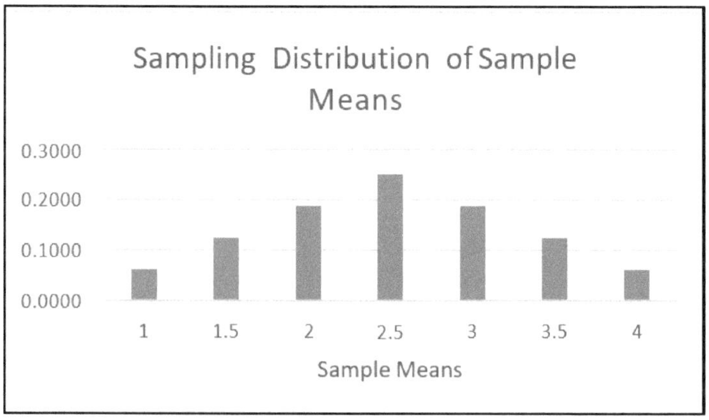

The results seen above are not surprising to statisticians. They have developed something called the

Central Limit Theorem that predicts that any population, regardless of shape, will result in a sampling distribution that is approximately normal. The mean of the sampling distribution, $\mu_{\bar{x}}$ will be the same as the mean of the population, and the standard deviation of the sampling distribution, $\sigma_{\overline{X}}$, also called the standard error of the mean, equals the population standard deviation divided by the square root of the sample size. Symbolically:

- $\mu_{\bar{x}} = \mu$
- $\sigma_{\bar{x}} = \frac{\sigma}{\sqrt{n}}$
- If n > 30, the sampling distribution will be normally distributed regardless of the shape of the population. If the population is normal, the sampling distribution will be normal for sampling distributions, even if n< 30.
- And, since we have the results in the first two bullets above, the z-score formula changes as follows:

$$z = \frac{\bar{x} - \mu_{\bar{x}}}{\sigma_{\bar{x}}} = \frac{\bar{x} - \mu}{\sigma / \sqrt{n}}$$

Use this formula whenever you are working with the mean of a ***sample of size n***. Our original z-score formula is used ***only*** when working with ***individual (non-sample) values of x.***

Example Suppose the diameter of all golf balls produced by a particular company is 1.682 inches (per the US Golf Association) with a standard deviation of 0.03 inches. If a sample of 36 balls is randomly selected from production to test adherence to the USGA specifications, what is the probability that the sample mean is less than 1.68 inches?

Solution You must be able to understand what is given in the problem. First, 1.682 is the mean of all golf balls, so μ = 1.682 inches; the standard deviation of these balls is 0.03 inches, so σ=0.03 inches. The sample size is n=36. The question is what is $P(\overline{x} < 1.68\,in)$?

You need to standardize the sample mean. So

$$z = \frac{\bar{x} - \mu_{\bar{x}}}{\sigma_{\bar{x}}} = \frac{1.68 - 1.682}{0.03 / \sqrt{36}} = \frac{-0.002}{0.005} = -0.4$$

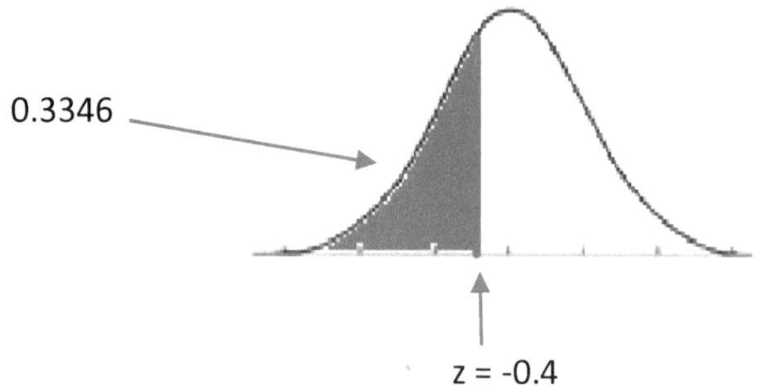

Therefore, the question is: $P(\bar{x} < 1.68\ in). = P(z < -0.4)$. Using a standard normal table, the cumulative area from $-\infty$ to z = -0.4 is 0.3346. Thus there is a 33.46% chance that samples of size 36 of the company's golf balls will have a diameter less than 1.68 inches. Seems too high – they may want to recalibrate!

Another statistic we can find is the sample proportion, $\hat{p}$ (read "p-hat"). As you recall, proportions are the result of the answer to a yes/no (binomial) question. We can also find the sampling distribution of sample proportions, just as we did for the sampling distribution of sample means. Since these proportions result from the response to binomial situations, $\hat{p} = \frac{x}{n}$, where x is the number of "successes" and n is the sample size. And since the standard error of the mean is related to the sample size, so is the standard error of the proportion: $\sigma_{\hat{p}} = \frac{\sigma}{\sqrt{n}}$. Therefore, $\sigma_{\hat{p}} = \sqrt{\frac{p(1-p)}{n}}$, where p is the population proportion and n is the sample size. [Please note: again, most texts (and your TI as well as StatCrunch) use $\hat{p}$ as the sample proportion and p as the population proportion; your homework in MyStatLab.com will use p for the sample proportion and π as the population proportion. Be careful!]

Consequently, our z-score formula is affected and must be modified: $z = \frac{\hat{p}-p}{\sqrt{\frac{p(1-p)}{n}}}$.

Example Suppose the national percentage of incoming community college students who need developmental coursework (at least one zero-level course) is 62%.
Recently, there was a sample of 60 new BCC students attending a new student orientation session. What is the probability that fewer than 65% needed developmental coursework? More than 65%? Between 62% and 65%?

Solution The problem tells us that nationally 62% of community college students need developmental courses, so p = 0.62 (be sure to change percents to decimals!) and (1 – p) = 0.38. The sample is n = 60 students, and are being asked about different situations involving the sample proportion from the sample, the first being $P(\hat{p} < 0.65)$. So:

$$z = \frac{\hat{p}-p}{\sqrt{\frac{p(1-p)}{n}}} = \frac{0.65-0.62}{\sqrt{\frac{0.62(1-0.62)}{60}}} = \frac{0.03}{\sqrt{0.003927}} \approx \mathbf{0.48}$$

Thus, $P(\hat{p} < 0.65) = P(z < 0.48) = 0.6844$. This means that, for samples of size 60, there is a 68.44% chance that fewer than 65% of those students will require developmental coursework.

The second question is the complement of the first one, so $P(\hat{p} > 0.65) = 1 - 0.6844 = 0.3156$, so there is a 31.56% chance that more than 65% of these students will need developmental coursework.

For the third question, notice that 62% is p, so its z-score is zero; hence, $P(0.62 < \hat{p} < 0.65) = P(0 < z < 0.48) = 0.6844 - 0.5000 = 0.1844$. This means there is an 18.44% probability that, when looking at samples of size 60, between 62% and 65% of those students will need developmental coursework.

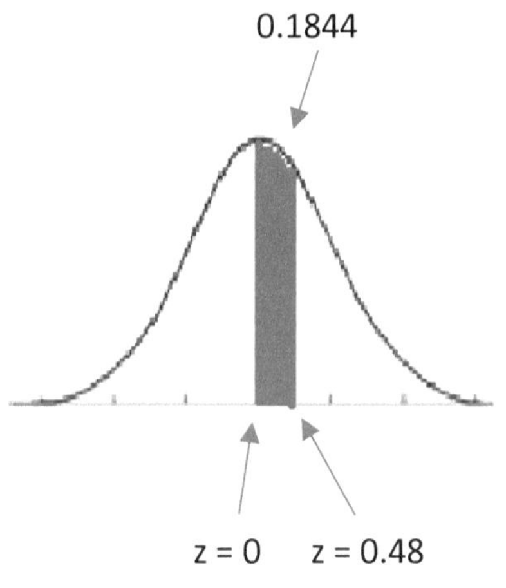

Confidence Intervals
Chapter 8 Summary

The Idea

Business analysts and others are often interested in finding out current public opinion or market trends. Since, as discussed in earlier chapters, it is impractical to survey the entire population of interest, statisticians often take samples and use the resulting statistics to make projections about the population. And, if the sample is chosen properly, the results will be very close to the actual population parameters in question. These sample statistics are referred to as *point estimates* of the respective population parameters.

However, it is highly unlikely that the sample statistic from a single sample will exactly equal the value of the population parameter, but it should be within a "reasonable" distance of the actual population parameter. In order to "compensate," statisticians construct *confidence intervals* about the sample statistic as a way of providing an interval estimate of the population parameter. These confidence intervals provide a "buffer zone" about the sample statistics that will, with high probability, contain the true value of the population parameter in question. For example, you may hear that a political poll states that a particular candidate has 48% of the vote. They may state that the "margin of error" is ±3%. What they have provided is a confidence interval of the candidate's share of the vote: 48% ± 3% means that the candidate's true proportion (p) of the vote lies somewhere between (48 – 3)% and (48 + 3)%, or from 45% to 51%, although we typically write proportions as decimal numbers. So, another way to state this confidence interval is as follows: $0.45 \leq p \leq 0.51$.

Depending on the population parameter to be estimated and the assumptions made, there are several different techniques to be utilized. They are summarized below.

<u>Estimating the Mean (σ known)</u>

The basic set-up for any confidence interval is: ***point estimate ± margin of error.***

The point estimate is the sample statistic (e.g., $\bar{x}$ or p). The margin of error is found by considering the level of confidence you want to have as well as the standard error of the sampling distribution. The size of the margin of error depends on the level of confidence – that is, do you want to be *somewhat* certain the population parameter is within your confidence interval, or *very* certain it is there. Typically, we use 90%, 95% or 99% confidence levels. The level of confidence determines the z-score you use in your calculations.

Elaine A. Previte
May 2019

All this sounds technical, and it is, but can be summarized as follows:

point estimate ± margin of error

$$\bar{x} \pm z\alpha/_2 \cdot \frac{\sigma}{\sqrt{n}}$$

Alpha (α) is the complement of the confidence level. So, if you're working with 95% confidence, then α = 1 - 0.95 = 0.05, and $z\alpha/_2$ is the z-score associated with $\alpha/_2$ area in each tail.

Here are some of the "most popular" z-sores you will be using:

Confidence Level	z-score
90%	1.645
95%	1.96
99%	2.575

Example Suppose you sample 36 of your fellow business majors and ask them how many hours per week they study. The sample mean is 15 hours, and the population standard deviation is known to be 2 hours. Construct a 95% confidence interval estimate of the true population mean number of hours business majors spend studying.

Solution Since we want a 95% confidence interval and we know that σ = 2, we willuse the formula given above with a z-score of 1.96. We also know that $\bar{x} = 15$. Thus,

point estimate ± margin of error

$$\bar{x} \pm z\alpha/_2 \cdot \frac{\sigma}{\sqrt{n}}$$

$$15 \pm (1.96) \cdot \frac{2}{\sqrt{36}}$$

$$15 \pm 0.6533$$

We can interpret this result as saying we are 95% confident that the true value of μ is somewhere in that interval; i.e., 15 – 0.6533 < μ < 15 + 0.6533 or 14.35 hr < μ < 15.65 hr. We can also say that μ will be found in 95% of such intervals.

Example Compute a 90% confidence interval for μ given the information above.

Solution Since we want a 90% confidence interval and we know that σ = 2, we will use the formula given above with a z-score of 1.645. We also know that $\bar{x} = 15$. Thus,

$$\bar{x} \pm z\alpha/_2 \cdot \frac{\sigma}{\sqrt{n}}$$

$$15 \pm (1.645) \cdot \frac{2}{\sqrt{36}}$$

$$15 \pm 0.5483$$

$$15 \pm 0.55$$

We are 90% confident that the true value of μ is somewhere in that interval; i.e., 15 – 0.55 < μ < 15 + 0.55 or 14.45 hr < μ < 15.55 hr.

Notice that a lower level of confidence yields a smaller margin of error than that found when using a higher level of confidence. Why??

What if I don't know σ?

In the real world, you typically don't know the value of the population standard deviation, σ. If you did know σ, then you would probably also know μ, so why would you need a confidence interval?

If you don't know σ, you cannot use a z-score when calculating a confidence interval. In this case, you need to use a new statistic, the t-score. The t-score is highly dependent on the sample size and confidence level, so we cannot give you a list of "popular" t-scores as we did z-scores. When looking up the t-score required for a particular confidence interval, you need to locate the *degrees of freedom* (df), equal to the sample size minus 1 (df = n -1) and the confidence level. So, for example, if you're looking for a 90% confidence interval for a sample of size 20 (hence df = 19), you look at the intersection of the df = 19 row and the 90% CI column. You can see the appropriate t-score is 1.729.

Elaine A. Previte
May 2019

Table F The *t* Distribution

d.f.	Confidence intervals	80%	90%	95%	98%	99%
	One tail, α	0.10	0.05	0.025	0.01	0.005
	Two tails, α	0.20	0.10	0.05	0.02	0.01
1		3.078	6.314	12.706	31.821	63.657
2		1.886	2.920	4.303	6.965	9.925
3		1.638	2.353	3.182	4.541	5.841
4		1.533	2.132	2.776	3.747	4.604
5		1.476	2.015	2.571	3.365	4.032
6		1.440	1.943	2.447	3.143	3.707
7		1.415	1.895	2.365	2.998	3.499
8		1.397	1.860	2.306	2.896	3.355
9		1.383	1.833	2.262	2.821	3.250
10		1.372	1.812	2.228	2.764	3.169
11		1.363	1.796	2.201	2.718	3.106
12		1.356	1.782	2.179	2.681	3.055
13		1.350	1.771	2.160	2.650	3.012
14		1.345	1.761	2.145	2.624	2.977
15		1.341	1.753	2.131	2.602	2.947
16		1.337	1.746	2.120	2.583	2.921
17		1.333	1.740	2.110	2.567	2.898
18		1.330	1.734	2.101	2.552	2.878
19		1.328	1.729	2.093	2.539	2.861
20		1.325	1.725	2.086	2.528	2.845
21		1.323	1.721	2.080	2.518	2.831
22		1.321	1.717	2.074	2.508	2.819

Example A sample of 20 incoming freshmen averages 22 hours per week of employment, with a sample standard deviation of 3 hours per week. Construct a 90% confidence interval estimate of the true mean number of hours per week of outside employment.

Solution Since we don't know the population standard deviation, σ, but we do know the sample standard deviation, s, we will use the t-distribution to help up construct the confidence interval.

Point Estimate ± Margin of Error

$$\bar{x} \pm t \cdot \frac{s}{\sqrt{n}}$$

$$22 \pm 1.729 \cdot \frac{3}{\sqrt{20}}$$

$$22 \pm 1.16$$

We are 90% confident that the true value of μ lies in the interval 20.84 < μ < 23.16 hours per week.

Elaine A. Previte
May 2019

Example Given the problem above, construct a 95% confidence interval estimate of the true mean number of hours per week of outside employment.

Solution This is the same process as above, but this time look under the 95% column. You will find a t-score of 2.093. So, the confidence interval is:

point estimate ± margin of error

$$\bar{x} \pm t \cdot \frac{s}{\sqrt{n}}$$

$$22 \pm 2.093 \cdot \frac{3}{\sqrt{20}}$$

$$22 \pm 1.404$$

We are 95% confident that the true value of μ lies in the interval 20.596 < μ < 23.404 hours per week.

Estimating the Population Proportion, *p*

Whenever we need to estimate a population proportion (p), we use the z-statistic. The process of finding a confidence interval for p is similar to the process we have already seen: *point estimate ± margin of error.* In this case, the point estimate for p is $\hat{p}$ (read "p-hat") and the standard error is $\sqrt{\frac{\hat{p}(1-\hat{p})}{n}}$. Use the z-score that corresponds to ${}^{\alpha}/_{2}$.

Example A sample of 200 students is taken to determine the proportion of the student body that is displeased with the college's cafeteria. 84 students responded that they were displeased. Find the 95% confidence interval for the true proportion of the college's students who are displeased with the cafeteria.

Solution Here x = 84, n = 200. Therefore $\hat{p} = \frac{84}{200} = 0.42$ and $1 - \hat{p} = 0.58$. The z-score to be used here is 1.96. Thus,

point estimate ± margin of error

$$\hat{p} \pm z{\alpha/_2} \cdot \sqrt{\frac{\hat{p}(1-\hat{p})}{n}}$$

$$0.42 \pm 1.96 \cdot \sqrt{\frac{(0.42)(0.58)}{200}}$$

$$0.42 \pm 1.96 \cdot 0.348998567$$

$$0.42 \pm 0.0684$$

Therefore, we are 95% confident that the true proportion of college students displeased with the cafeteria is between 0.3516 and 0.4884 (35.16% to 48.84%).

Sample Size

Sometimes a researcher wants to have a particular margin of error. For example, if a company is conducting political polls for an upcoming election, they will not want to have a large margin of error (traditionally, you see errors such as ± 3% or ± 3.5%; larger margins might provide meaningless results). Since margin of error is related to sample size, the researcher wants to find a sample large enough to ensure that the margin of error is appropriate for the situation under study.

Sample Size for Estimates of the Population Mean, μ

Example A university dean wants to estimate the mean number of hours students spend studying each week. The standard deviation from a previous study is known to be 2.6 hours. How large a sample should be taken to estimate the mean within 0.5 hours with 95% confidence.

Solution Recall that the margin of error, E, is defined as $z_{\alpha/2} \cdot \frac{\sigma}{\sqrt{n}}$. In this problem, we need E to be 0.5 hr; σ = 2.6 hr; and the z-score is 1.96. Solving for n:

$$E = z_{\alpha/2} \cdot \frac{\sigma}{\sqrt{n}}$$

$$(E)\sqrt{n} = \left(z_{\alpha/2} \cdot \frac{\sigma}{\sqrt{n}}\right)\sqrt{n}$$

$$\frac{(E)\sqrt{n}}{E} = \frac{z_{\alpha/2} \cdot \sigma}{E}$$

$$\sqrt{n} = \frac{z_{\alpha/2} \cdot \sigma}{E}$$

$$\left(\sqrt{n}\right)^2 = \left(\frac{z_{\alpha/2} \cdot \sigma}{E}\right)^2$$

$$n = \left(\frac{z_{\alpha/2} \cdot \sigma}{E}\right)^2$$

So, using the result, we get:

$$n = \left(\frac{z_{\alpha/2} \cdot \sigma}{E}\right)^2 = \left(\frac{1.96 \cdot 2.6}{0.5}\right)^2 = (10.192)^2 = 103.87 \rightarrow 104$$

We must always round up to the next whole number (since we are looking at how many students we need to sample). Thus, the minimum sample size required to meet the margin of error is 104.

Sample Size for Estimates of the Population Proportion, *p*

Example A researcher wishes to estimate, with 90% confidence, the true proportion of households that own a computer or tablet. A previous study showed that 40% of households owned one of these devices. How large a sample must be taken to be within 2% of the true proportion? (Note: change 2% to 0.02.)

Solution We know that the margin of error in this case is

$$E = z_{\alpha/2} \cdot \sqrt{\frac{\hat{p}(1-\hat{p})}{n}}.$$

Using basic algebra, we get

$$n = \hat{p}(1-\hat{p})\left(\frac{z_{\alpha/2}}{E}\right)^2.$$

Evaluating, we get

$$n = \hat{p}(1-\hat{p})\left(\frac{z_{\alpha/2}}{E}\right)^2 = (0.40)(0.60)\left(\frac{1.645}{0.02}\right)^2 = 1623.615 \rightarrow 1624 \quad .$$

Note: If a sample proportion is not given, use $\hat{p} = 0.5$. This value will ensure a sample size that will meet the required margin of error.

Hypothesis Testing – One Sample
Chapter 9 Summary

The Idea

In business and economics, people often make assumptions (hypotheses) about the value of a population parameter like μ or π (or p in most texts). Their assumptions may or may not be accurate, so they routinely take random samples and use those sample statistics to test their hypotheses. If their sample's z-score (or t-score) results are "reasonably" close to the hypothesized value of the parameter, then they cannot reject their hypothesis (the null hypothesis – see below). If, on the other hand, their sample's z-score (or t-score) results are not "reasonably" close to the hypothesized value of the parameter, then one can infer that it is highly unlikely to encounter such a result if, in fact, the hypothesized value of the parameter is correct. In this case, they would reject the null hypothesis.

If all this sounds overwhelming, it is understandable. However, once you understand the process, it becomes a relatively easy task to accomplish. First, we will start with μ, with σ known. In this case, then, we will use the z-statistic in our calculations.

Let us make some basic definitions before we begin.

Term	**Definition**
H_0	Null Hypothesis – the statement that a population parameter equals a particular value
H_1	Alternative Hypothesis – the statement that is the exact mathematical opposite of H_0
α	Alpha, the Level of Significance – this is the probability of rejecting a true null hypothesis (H_0); you want to keep this value small (e.g., 0.01, 0.05, 0.10, but any value between 0 and 1 can be used); $1 - \alpha$ is the confidence level
Test Statistic	The value of the statistic to be used in the hypothesis test (in this case, z)
Critical Value	The value of the z-statistic (or other statistic as we go forward) that will partition the rejection region from the non-rejection region under the normal curve
Rejection Region	That area under the normal curve that is in the tail(s); should your test statistic be in the rejection region you will reject H_0
Non-rejection Region	That area under the normal curve that the complement of the rejection region; should your test statistic be in the non-rejection region you will fail to reject H_0
p-value	The probability of the test statistic being beyond the value of the test statistic going toward ∞ or $-\infty$.

Classical Hypothesis Testing for μ (σ known)

There is a five-step process in classical hypotheses testing. Each will be defined below, followed by two examples.

Step 1: State the Hypotheses

In hypothesis testing, there are two hypotheses: the null hypothesis, denoted H_0, and the alternative hypothesis, H_a or H_1 (depending on the textbook). As shown in the table above, the null hypothesis ***must*** contain a statement of equality about the population parameter; for example, $\mu = 20$ or $\mu \leq 100$. The value of μ will be used when we do our calculations, so we need it to have a given value. Often, but not always, the value someone is claiming about μ will be stated as the null hypothesis.

The alternative hypothesis, H_1, contains the mathematical statement that is the *exact opposite* of H_0. See the table below.

	Case 1	Case 2	Case 3
H_0	$\mu = 20$	$\mu \leq 20$	$\mu \geq 20$
H_1	$\mu \neq 20$	$\mu > 20$	$\mu < 20$
Number of Tails	2	1 (upper, or right, tail)	1 (lower, or left, tail)
Illustration	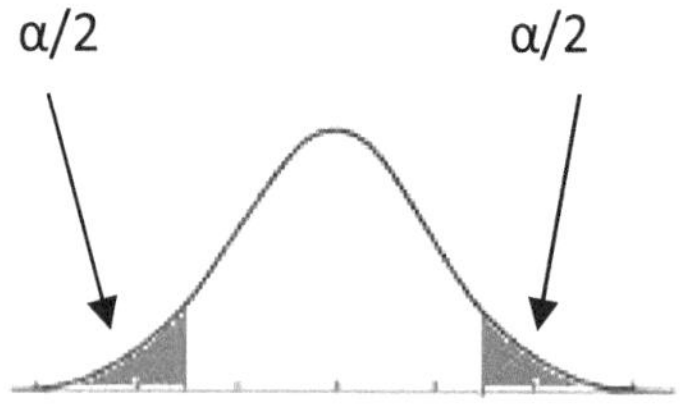	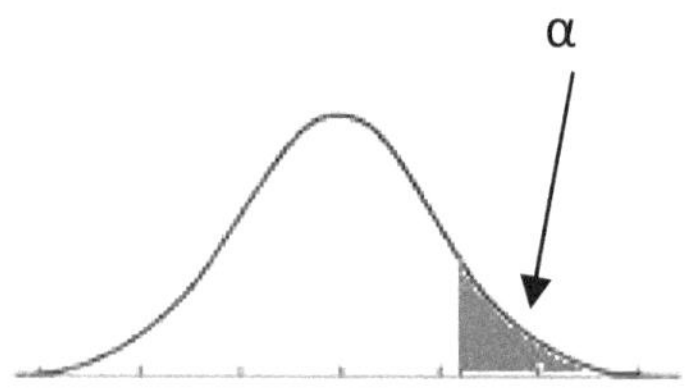	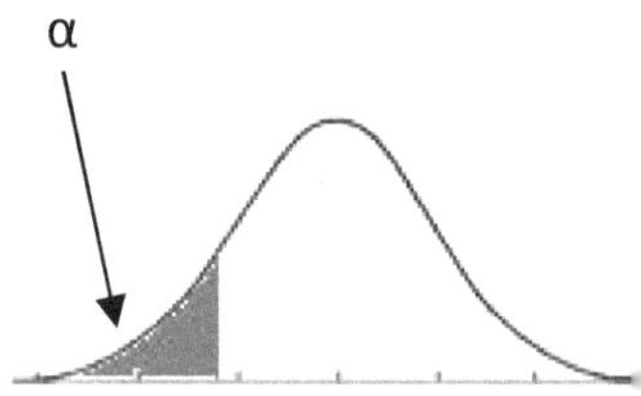

Step 2: Locate the Rejection Region

The rejection region is that area under the curve from a critical value to ∞ (or $-\infty$) and depends on the α level. The critical value of the z-statistic (in this case) is that z-score such that the appropriate amount of area under the curve is located from $-\infty$ to the critical z-score, or from the critical z-score to ∞. If the test statistic's value falls in the rejection region, we will reject the null hypothesis; if not, we will fail to reject the null hypothesis. (Note that we do not say we "accept" the null hypothesis.)

Step 3: Calculate the Test Statistic

If we are testing for μ and σ is known, we will use the z-statistic. Underlying assumptions are that the population is normal or nearly normal, and that simple random samples were taken.

If we are testing for μ and σ is *un*known, we will use the t-statistic. We will cover this scenario later.

Elaine A. Previte
May 2019

Step 4: **Make a Decision about H_0**

See where the value of the test statistic lies relative to the critical z-score. If the test statistic falls in the rejection region, we reject H_0; otherwise we fail to reject H_0. Note: we never say we "accept" H_0, just as a defendant is never found "innocent" in a court proceeding: we just haven't yet found evidence to reject H_0 (or convict the defendant).

Step 5: **State your Conclusion**

Once you have made your decision about H_0, you should relate it back to the context of the problem.

Example 1 **GoGo Cola**

GoGo Cola, a local soda company fills two-liter bottles with a mean of 2.0 liters. It is known that the standard deviation of the filled bottles is 0.2 L. The production manager wishes to test the filling process to ensure that the machinery is working properly. She selects a random sample of 64 bottles and finds that the sample mean is 1.95 L. At the 0.05 level of significance, does this support the company's contention that the mean amount of Cola is 2.0 L.?

Solution

Step 1: **State the Hypotheses**

Since the question is whether μ ***is*** 2.0 L., this becomes our null hypothesis. The alternative hypothesis is that μ ***is not*** 2.0 L. We typically write:

H_0: $\mu = 2.0$ L. (this is the claim) H_1: $\mu \neq 2.0$ L.

Step 2: **Locate the Rejection Region**

The rejection region is determined by the alternative hypothesis. Since H_1 states that μ does not equal 2.0 L., this can happen if μ is more than 2.0 or if μ is less than 2.0. This means that we have a two-tailed test.

A significance level of 0.05 in a two-tailed test means we must divide the .05 area between the two tails: 0.025 area in each tail. This corresponds to critical z-scores of $\pm$ 1.96.

We will reject H_0 if our test value is either less than -1.96 or greater than 1.96.

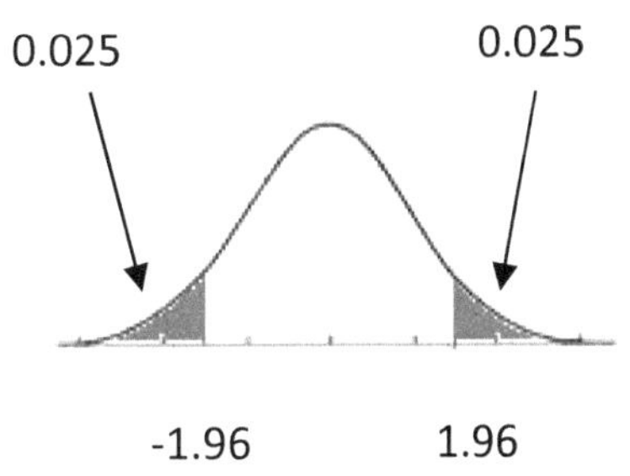

Step 3: **Calculate the Test Statistic**

We know:	Thus:
$\mu = 2.0$ $\sigma = 0.2$ $n = 64$ $\bar{x} = 1.95$	$$z = \frac{\bar{x} - \mu}{\sigma / \sqrt{n}} = \frac{1.95 - 2.0}{0.2 / \sqrt{64}} = \frac{-0.05}{0.025} = -2.0$$ 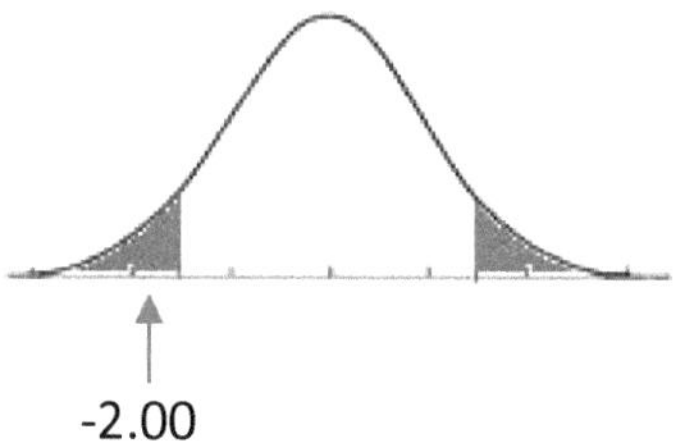

Step 4: **Make a Decision about H_0**

Since $-2.00 < -1.96$, the test statistic falls in the rejection region. We reject H_0.

Step 5: **State your Conclusion**

At the 0.05 level of significance, there is sufficient evidence to reject the company's claim that the mean fill amount is 2.0 L.

Example 2 **Venti-Trenti Coffee**

Venti-Trenti Coffee, a very famous coffee company, sells medium iced coffees that contain, on average, 20 fl. oz. of coffee. They state that the population standard deviation is 0.8 fl. oz. A consumer advocacy group believes the coffee company under-filling their medium ice coffees, so they take a sample of 100 medium ice coffees and find the sample mean to be 19.9 fl. oz. Using $\alpha = 0.05$, is the consumer group's claim correct?

Solution

Step 1: **State the hypotheses**

Since the consumer group claims the company is under-filling the medium iced coffees, the statement $\mu < 20$ goes in the alternative hypothesis. The null hypothesis contains the opposite statement, $\mu \geq 20$.

H_0: $\mu \geq 20$ fl. oz.

H_1: $\mu < 20$ fl. oz.   (this is the claim)

Step 2: **Locate the Rejection Region**

The rejection region is determined by the alternative hypothesis. Since H_1 states that $\mu < 20$, we have a one-tailed test.

A significance level of 0.05 in a one-tailed test corresponds to critical z-score of -1.645.

We will reject H_0 if our test value is less than -1.645.

$\alpha = 0.05$

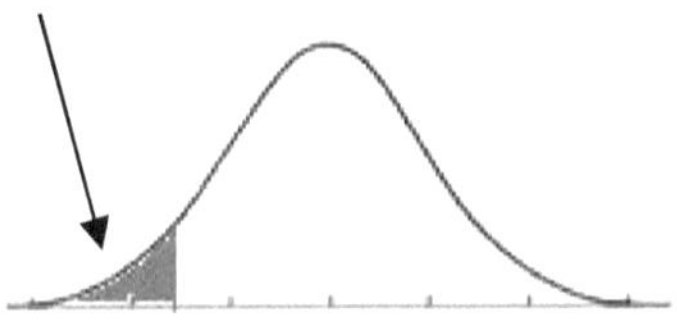

-1.645

Step 3: **Calculate the Test Statistic**

We know:	Thus:
$\mu = 20$ $\sigma = 0.8$ $n = 100$ $\bar{x} = 19.9$	$z = \dfrac{\bar{x} - \mu}{\sigma / \sqrt{n}} = \dfrac{19.9 - 20.0}{0.8 / \sqrt{100}} = \dfrac{-0.10}{0.08} = -1.25$ 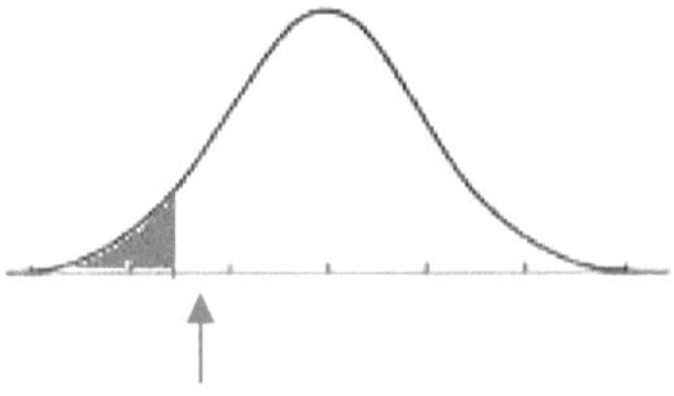 -1.25

Step 4: **Make a Decision about H_0**

Since -1.25 > -1. 645, the test statistic does *not* fall in the rejection region. We *fail to reject* H_0.

Step 5: **State your Conclusion**

At the 0.05 level of significance, there is insufficient evidence to support the consumer group's claim that the medium iced coffees are under-filled.

p-Value Approach to Hypothesis Testing

A more modern approach to hypothesis testing involves the comparison of the test statistic's p- value to the α-level. If the p-value is less than the α-level, this means that the test statistic falls in the rejection region. If the p-value is greater than the α-level, the test statistic does not fall in the rejection region. A simple way to remember this is: "If p is low, H_0 must go."

To find the p-value, simply find the area from the test statistic to ∞ or - ∞, depending on the alternative hypothesis. We will look at the previous examples in reverse order.

Example 2 Venti-Trenti Coffee

The value of the test statistic here was z = -1.25. The area from - ∞ to -1.25 can be found by looking at the standard normal table. Here the p-value is 0.1056. Since 0.1056 > 0.05, we fail to reject H_0.

Example 1 GoGo Cola

The value of the test statistic here was z = -2.0, so the area is 0.0228. Since this was a two-tailed test, we must double this area to obtain the p-value. So, the p-value equals 2(0.0228) = 0.0556, which is greater than the α-level of 0.05. Thus, we fail to reject the null hypothesis.

Hypothesis Testing for μ (σ unknown)

In the case where σ is unknown (which is most of the time!), we need to use the t-statistic when performing hypothesis tests. Other than that change, the process is the same as described above.

Example 3 Online Savings

A marketing company specializing in online coupons tries to recruit new retail customers by boasting an average increase in savings of more than $35 per transaction. To corroborate their claim, they randomly select a sample of 20 transactions and find the mean savings to be $38 with a standard deviation of $5. At the 0.10 level of significance, test whether the marketing company's claim is correct.

Solution

Step 1: State the hypotheses

Since the marketing company claims they save customers an average of more than $35, i.e., $\mu > 35$, this statement must go in the alternative hypothesis. The null hypothesis is $\mu \leq 35$.

H_0: μ ≤ \$35

H_1: μ > \$35 (this is the claim)

Step 2: **Locate the Rejection Region**

The rejection region is determined by the alternative hypothesis. Since H_1 states that μ > 35, we have an upper tail test.

A significance level of 0.10 with df = 19 in a one-tailed test corresponds to critical t-score of 1.328

We will reject H_0 if our test value is greater than 1.328.

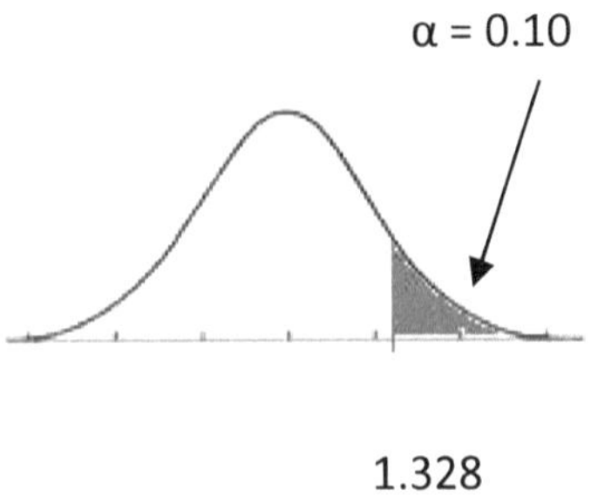

Step 3: **Calculate the Test Statistic**

We know:	Thus:
μ = 35 s = 5 n = 20 df = 19 $\bar{x}$= 38	$t = \frac{\bar{x}-\mu}{s/\sqrt{n}} = \frac{38-35}{5/\sqrt{20}} = \frac{3}{1.11803} = 2.683$ 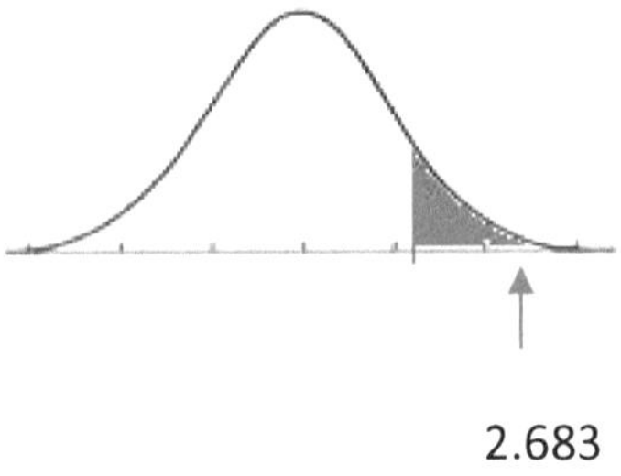

Step 4: **Make a Decision about H_0**

Since 2.683 > 1. 328, the test statistic falls in the rejection region. We *reject* H_0.

Step 5: **State your Conclusion**

At the 0.10 level of significance, there is sufficient evidence to support the company's claim that the mean savings are more than \$35.

Elaine A. Previte
May 2019

You can see the process is the same regardless of the test statistic used. However, finding the p- value for the t-statistic is more difficult than for the z-statistic. You must use technology (the TI or *StatCrunch*) to assist. See the TI step-by-step directions for finding the p-value in this case (use the inverse t-distribution).

Hypothesis Testing for a Population Proportion

We can also conduct hypothesis tests for population proportions. The process is similar to what we have already done; but in all cases, we use the z statistic when dealing with population proportions.

Example 4 Student Loan Debt

In 2012, 66% of students who graduated from public colleges and universities had student loan debt (data from TICAS.org). You want to see if this is true for BCC students, so you take a random sample of 200 graduating BCC students and find that 120 of them have student loan debt. At the 0.05 level, does your data support the claim that BCC has a lower percentage of graduating students with college loan debt than the national average?

Solution

Step 1: State the hypotheses

Since this question involves percentages (proportions), we are testing for the population proportion, p (not μ!).

H_0: $p \geq 0.66$

H_1: $p < 0.66$ (this is the claim)

Step 2: Locate the Rejection Region

A significance level of 0.05 in a left-tailed test corresponds to critical z-score of -1.645.

We will reject H_0 if our test value is less than -1.645.

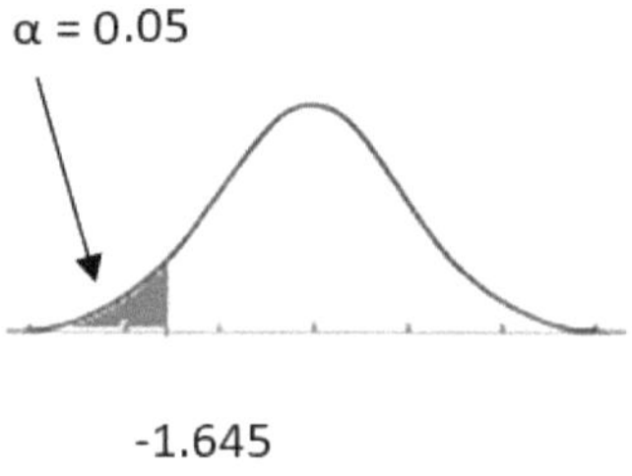

Step 3: **Calculate the Test Statistic**

We know:	Thus:
p = 0.66 q = 1 – p = 1 – 0.66 = 0.34 n = 200 x = 120 $\hat{p} = \frac{x}{n} = \frac{120}{200} = 0.60$	$z = \frac{\hat{p} - p}{\sqrt{\frac{pq}{n}}} = \frac{.60 - .66}{\sqrt{\frac{(.66).(34)}{200}}} = \frac{-.06}{0.033496} = -1.79$ 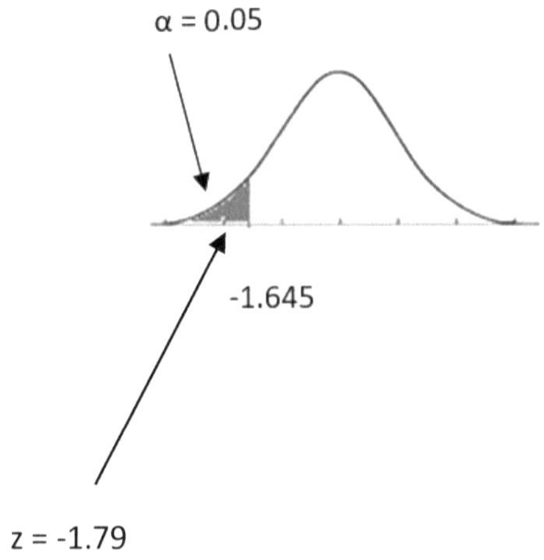

Step 4: **Make a Decision about H_0**

Since -1.79 < -1.645, the test statistic falls in the rejection region. We *reject* H_0.

Step 5: **State your Conclusion**

At the 0.05 level of significance, there is sufficient evidence to support the claim that a lower proportion of graduating BCC students have student loan debt.

Hypothesis Testing – Two Samples
Chapter 10 Summary

The Idea

In the previous chapter, we conducted hypothesis tests for population means and proportions by taking one sample. When testing for population means, we used the z-statistic when σ is known; we used the t-statistic when σ is unknown. When testing for population proportions, we used the z-statistic.

There are many situations in which researchers need to know the difference between *two* populations. For example, when testing a new drug, researchers will have an experimental group (those who take the new drug) and a control group (those who take a placebo); they are interested to see if any changes exhibited are statistically significant or simply due to sampling error.

We will examine the following situations:

- The difference between two independent population means when σ is known (z-stat)
- The difference between two independent population means when σ is unknown (t-stat)
    - If the two population variances are equal, use a pooled t-test
    - If the two population variances are unequal, use a separate-variance t-test
- The difference between two dependent population means when σ is unknown (t-stat)
- The difference between two independent population variances (F-stat)

It is important to note that in each of the above-mentioned situations we will use variations of the formulas we used in Chapter 9.

Recall our basic definitions:

Term	**Definition**
H_0	Null Hypothesis – the statement that a population parameter equals a particular value
H_1	Alternative Hypothesis – the statement that is the exact mathematical opposite of H_0
α	Alpha, the Level of Significance – this is the probability of rejecting a true null hypothesis (H_0); you want to keep this value small (e.g., 0.01, 0.05, 0.10, but any value between 0 and 1 can be used); 1 – α is the confidence level
Test Statistic	The value of the statistic to be used in the hypothesis test (in this case, z)
Critical Value	The value of the z-, t- or F-statistic that will partition the rejection region from the non-rejection region under the normal curve
Rejection Region	That area under the normal curve that is in the tail(s); should your test statistic be in the rejection region you will reject H_0

Non-rejection Region	That area under the normal curve that the complement of the rejection region; should your test statistic be in the non-rejection region you will fail to reject H_0
p-value	The probability of the test statistic being beyond the value of the test statistic going toward ∞ or $-\infty$.

Classical Hypothesis Testing for the Difference Between Two Independent Population Means ($\sigma 1$ and $\sigma 2$ are known)

There is a five-step process in classical hypotheses testing. Each will be defined below, followed by two examples.

Step 1: State the Hypotheses

In hypothesis testing, there are two hypotheses: the null hypothesis, denoted H_0, and the alternative hypothesis, H_a or H_1 (depending on the textbook). Recall that the null hypothesis ***must*** contain a statement of equality about the population parameter.

Since we are looking at the difference between *two* means we have two means, μ_1 and μ_2. We are interested in whether these values are equal, or if there is another relationship between them. See the table below. Keep in mind that if $\mu_1 = \mu_2$, it can also be said that $\mu_1 - \mu_2 = 0$. This value will be used when we do our calculations. Note: it is possible, but not likely, that the difference between μ_1 and μ_2 will be some other specified value.

The alternative hypothesis, H_1, contains the mathematical statement that is the *exact opposite* of H_0. See the table below.

	Case 1	Case 2	Case 3
H_0	$\mu_1 = \mu_2$ or $\mu_1 - \mu_2 = 0$	$\mu_1 \le \mu_2$ or $\mu_1 - \mu_2 \le 0$	$\mu_1 \ge \mu_2$ or $\mu_1 - \mu_2 \ge 0$
H_1	$\mu_1 \ne \mu_2$ or $\mu_1 - \mu_2 \ne 0$	$\mu_1 > \mu_2$ or $\mu_1 - \mu_2 > 0$	$\mu_1 < \mu_2$ or $\mu_1 - \mu_2 < 0$
Number of Tails	2	1 (upper, or right, tail)	1 (lower, or left, tail)
Illustration	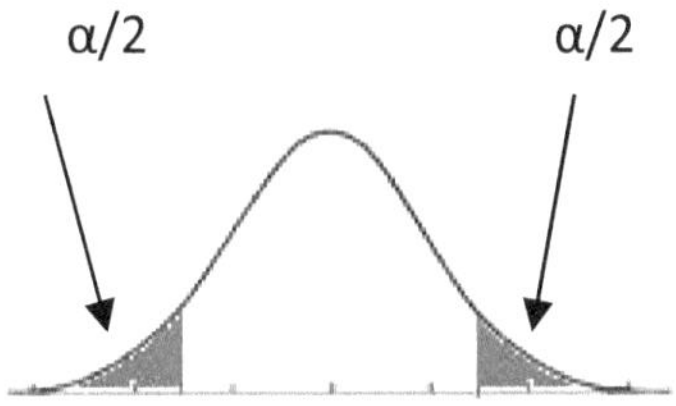	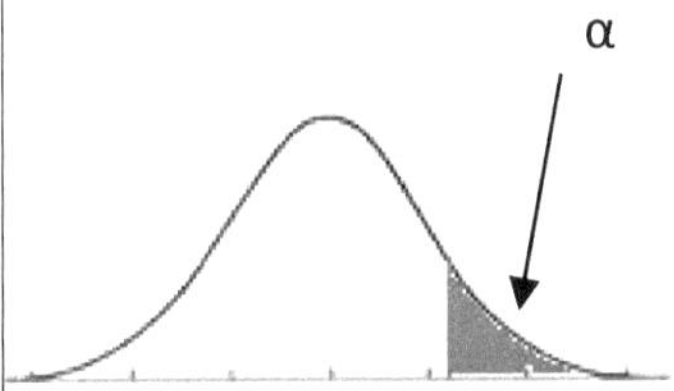	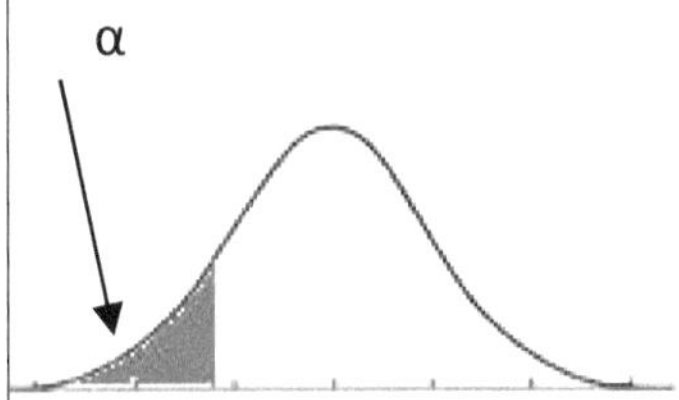

Step 2: Locate the Rejection Region

The rejection region is that area under the curve from a critical value to ∞ (or $-\infty$) and depends on the α level. The critical value of the z-s or t-statistic is that z- or t-score such that the appropriate amount

of area under the curve is located from $-\infty$ to the critical z- or t-score, or from the critical z- or t-score to ∞. If the test statistic's value falls in the rejection region, we will reject the null hypothesis; if not, we will fail to reject the null hypothesis. (Note that we do not say we "accept" the null hypothesis.)

Step 3: Calculate the Test Statistic

If we are testing for the difference between μ_1 and μ_2 and σ is known, we will use the z-statistic. Underlying assumptions are that the population is normal or nearly normal, and that simple random samples were taken. The formula we use is a modification of our original z-stat formula.

Our test statistic is:

$$z = \frac{(\bar{x}_1 - \bar{x}_2) - (\mu_1 - \mu_2)}{\sqrt{\frac{\sigma_1^2}{n_1} + \frac{\sigma_2^2}{n_2}}}$$

If we are testing for the difference between μ_1 and μ_2 and σ is *un*known, we will use the t-statistic. We will cover this scenario later.

Step 4: Make a Decision about H_0

See where the value of the test statistic lies relative to the critical z-score. If the test statistic falls in the rejection region, we reject H_0; otherwise we fail to reject H_0.

Step 5: State your Conclusion

Once you have made your decision about H_0, you should relate it back to the context of the problem.

Example: Credit Card Debt

In a recent survey of college students, the average credit card debt for a recent year was $9205. Five years earlier a similar survey found the average student credit card debt was $7618. Assume sample sizes of 35 were used and the population standard deviations of both samples were $1928. Is there enough evidence to believe that the average credit card debt has increased? Use $\alpha = 0.05$.

Solution Note: $\bar{x}_1$ = $9205 and $\bar{x}_2$ = $7618; $n_1 = n_2 = 35$; $\sigma_1 = \sigma_2 = 1928$

Step 1: State the Hypotheses

Since the question is whether there is a statistically significant increase in the difference between μ_1 and μ_2, this becomes our *alternative* hypothesis. The null hypothesis is that the difference between μ_1 and μ_2 has not increased (i.e., it has decreased or remained the same).

H_0: $\mu_1 \leq \mu_2$ (or $\mu_1 - \mu_2 \leq 0$)

H_1: $\mu_1 > \mu_2$ (or $\mu_1 - \mu_2 > 0$) (this is the claim)

Step 2: **Locate the Rejection Region**

The rejection region is determined by the alternative hypothesis. Since H_1 states that $\mu_1 > \mu_2$, we have a right-tailed test. The critical value of z is 1.645, so we will reject H_0 if our test value is greater than 1.645.

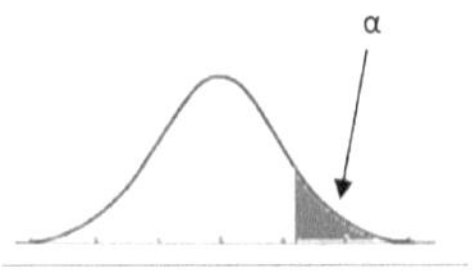

Step 3: **Calculate the Test Statistic**

We know:	Thus:
$\bar{x}_1$ = **\$9205** $\bar{x}_2$ = **\$7618** $n_1 = n_2$ = **35** $\sigma_1 = \sigma_2$ = **1928**	$z = \dfrac{(\bar{x}_1 - \bar{x}_2) - (\mu_1 - \mu_2)}{\sqrt{\dfrac{\sigma_1^2}{n_1} + \dfrac{\sigma_2^2}{n_2}}} = \dfrac{(9205 - 7618) - (0)}{\sqrt{\dfrac{1928^2}{35} + \dfrac{1928^2}{35}}} = \dfrac{1587}{\sqrt{212{,}410.51429}} = 3.443$

Step 4: **Make a Decision about H_0**

Since 3.443 > 1.645, the test statistic falls in the rejection region. We reject H_0.

Step 5: **State your Conclusion**

At the 0.05 level of significance, there is sufficient evidence to support the claim that the mean student credit card debt has increased.

Note: If we were doing this hypothesis test by the p-value method, we would find the area under the normal curve from z = 3.443 to ∞. The cumulative area from -∞ to 3.443 is 0.9997, so the p-value is 1 – 0.9997 = 0.0003. Since this value is less than the α-level (0.05), we reject the null hypothesis.

Classical Hypothesis Testing for the Difference Between Two Independent Population Means (σ_1 and σ_2 unknown)

Most often, σ_1 and σ_2 are unknown, so we cannot use the z-stat. However, we cannot simply use the t-stat: one must know about the variances of the populations involved. There are two separate possibilities: $\sigma_1^2 = \sigma_2^2$ or $\sigma_1^2 \neq \sigma_2^2$. At this point, the status of σ_1 and σ_2 will be given to you in the problem.

Case 1: $\sigma_1^2 = \sigma_2^2$

If the population variances are equal, we need to use a ***pooled-variance t-test*** to perform the hypothesis test for the difference between two population means. We will modify the traditional steps as we go through the following example.

Example: Lunch-Time Lines at the Bank

A bank with a branch located in the business district of a city has the objective of decreasing customer wait time during the noon-to-l P.M. lunch period. Wait time is the number of minutes that elapses from when the customer enters the line until he or she reaches the teller window. Data are collected from a random sample of 15 customers:

4.21	5.55	3.02	5.13	4.77	2.34	3.54	3.20
4.50	6.10	0.38	5.12	6.46	6.19	3.79	

Another branch is located in a residential area, and is also concerned with decreasing wait time during the noon-to-l P.M. lunch period. Data are collected from a random sample of 13 customers:

9.66	5.90	8.02	5.79	8.73	3.82	8.01
8.35	6.68	5.64	4.08	6.17	5.47	

Assuming that the population variances from both banks are ***equal,*** is there evidence of a difference in the mean waiting time between the two branches? (Use α = 0.05.)

Note: We need to compute the mean and sample variances of each sample.
City branch: $\bar{x}_1 = 4.28667;\ s_1^2 = 2.682995;\ n_1 = 15$
Residential branch: $\bar{x}_2 = 6.64;\ s_1^2 = 3.213447;\ n_2 = 13$

Step 1: **State the hypotheses**

Since the bank is interested in whether there is a statistically significant difference between the two means, the statement $\mu_1 \neq \mu_2$ or $\mu_1 - \mu_2 \neq 0$ goes in the alternative hypothesis. The null hypothesis contains the opposite statement, $\mu_1 = \mu_2$ or $\mu_1 - \mu_2 = 0$.

H_0: $\mu_1 - \mu_2 = 0$

H_1: $\mu_1 - \mu_2 \neq 0$ (this is the claim)

Step 2: **Locate the Rejection Region**

The rejection region is determined by the alternative hypothesis. Since H_1 states that $\mu_1 - \mu_2 \neq 0$, we have a two-tailed test. The number of degrees of freedom is $n_1 + n_2 - 2$; 26 df in this case,

A significance level of 0.05 in a two-tailed test with 26 df corresponds to critical t-score of $\pm$2.0555.

We will reject H_0 if our test value is less than -2.0555 or greater than 2.0555.

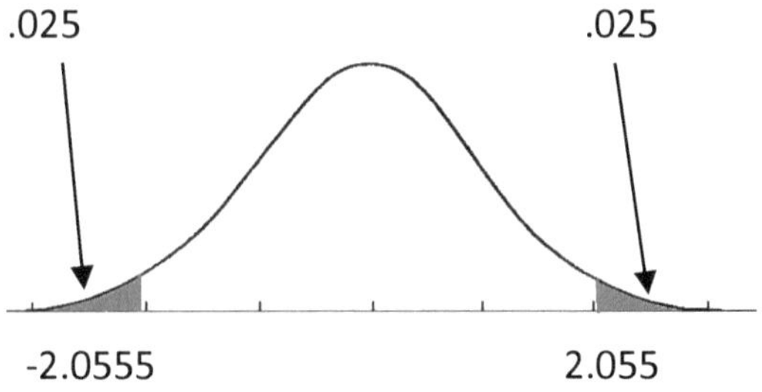

Step 3: **Calculate the Test Statistic**

We know the following:

City branch: $\bar{x}_1 = 4.28667;\ s_1^2 = 2.682995;\ n_1 = 15$
Residential branch: $\bar{x}_2 = 76.64\ s_2^2 = 3.213447;\ n_2 = 13$

We need to compute the ***pooled variance,*** s_p^2. This is basically a weighted variance.

$$s_p^2 = \frac{(n_1 - 1)s_1^2 + (n_2 - 1)s_2^2}{n_1 + n_2 - 2} = \frac{(15 - 1)(2.682995) + (13 - 1)(3.213447)}{15 + 13 - 2} = 2.92782$$

df = 15 + 13 – 2 = 26

The t-stat formula is modified for this situation:

$$t = \frac{(\bar{x}_1 - \bar{x}_2) - (\mu_1 - \mu_2)}{\sqrt{s_p^2\left(\frac{1}{n_2} + \frac{1}{n_2}\right)}} = \frac{(4.28667 - 6.64) - (0)}{\sqrt{2.92782\left(\frac{1}{15} + \frac{1}{13}\right)}} = \frac{-2.35333}{0.64839} = -3.6295$$

Step 4: **Make a Decision about H_0**

Since -3.6295 < -2.0484, the test statistic does fall in the rejection region. We reject H_0.

Step 5: **State your Conclusion**

At the 0.05 level of significance, there is sufficient evidence to reject the null hypothesis and support the bank's claim that there is a difference in mean wait times at the different bank branches.

Note: the p-value is 0.00122.

Case 2: $\sigma_1^2 \neq \sigma_2^2$

If the population variances are ***unequal***, we need to use a ***separate variance t-test*** to perform the hypothesis test for the difference between two population means. We will modify the traditional steps as we go through the Bank example.

Example: Lunch-Time Lines at the Bank

A bank with a branch located in the business district of a city has the objective of decreasing customer wait time during the noon-to-l P.M. lunch period. Wait time is the number of minutes that elapses from when the customer enters the line until he or she reaches the teller window. Data are collected from a random sample of 15 customers:

4.21	5.55	3.02	5.13	4.77	2.34	3.54	3.20
4.50	6.10	0.38	5.12	6.46	6.19	3.79	

Another branch is located in a residential area, and is also concerned with decreasing wait time during the noon-to-l P.M. lunch period. Data are collected from a random sample of 13 customers:

9.66	5.90	8.02	5.79	8.73	3.82	8.01
8.35	6.68	5.64	4.08	6.17	5.47	

Assuming that the population variances from both banks are ***unequal***, is there evidence of a difference in the mean waiting time between the two branches? (Use $\alpha = 0.05$.)

Note: We need to compute the mean and sample variances of each sample.

City branch: $\bar{x}_1 = 4.28667;\ s_1^2 = 2.682995;\ n_1 = 15$

Residential branch: $\bar{x}_2 = 6.64;\ s_2^2 = 3.213447;\ n_2 = 13$

Step 1: **State the hypotheses**

H₀: $\mu_1 - \mu_2 = 0$

H₁: $\mu_1 - \mu_2 \neq 0$ (this is the claim)

Step 2: **Locate the Rejection Region**

The rejection region is determined by the alternative hypothesis. Since H₁ states that $\mu_1 - \mu_2 \neq 0$, we have a two-tailed test. The number of degrees of freedom is computed by using the following formula:

$$df = \gamma = \frac{\left(\frac{s_1^2}{n_1} + \frac{s_2^2}{n_2}\right)^2}{\frac{\left(\frac{s_1^2}{n_1}\right)^2}{n_1 - 1} + \frac{\left(\frac{s_2^2}{n_2}\right)^2}{n_2 - 1}} = \frac{\left(\frac{2.682995}{15} + \frac{3.213447}{13}\right)^2}{\frac{\left(\frac{2.682995}{15}\right)^2}{14} + \frac{\left(\frac{3.213447}{13}\right)^2}{12}} = 24.606$$

A significance level of 0.05 in a two-tailed test with 24.606 df corresponds to critical t-score of ± 2.0612.

We will reject H₀ if our test value is less than -2.0612 or greater than 2.0612.

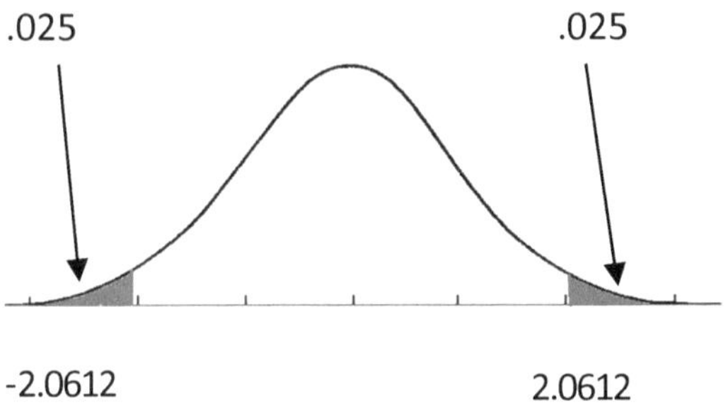

Step 3: **Calculate the Test Statistic**

We know the following:

City branch: $\bar{x}_1 = 4.28667;\ s_1^2 = 2.682995;\ n_1 = 15$

Residential branch: $\bar{x}_2 = 6.64;\ s_1^2 = 3.213447;\ n_2 = 13$

The t-stat formula is modified for this situation:

$$t = \frac{(\bar{x}_1 - \bar{x}_2) - (\mu_1 - \mu_2)}{\sqrt{\frac{s_1^2}{n_2} + \frac{s_2^2}{n_2}}} = \frac{(4.28667 - 6.64) - (0)}{\sqrt{\frac{2.682995}{15} + \frac{3.213447}{13}}} = \frac{-2.35333}{0.652729} = -3.60538$$

Step 4: Make a Decision about H_0

Since -3.6057 < -2.0612, the test statistic does fall in the rejection region. We reject H_0.

Step 5: State your Conclusion

At the 0.05 level of significance, there is sufficient evidence to reject the null hypothesis and support the bank's claim that there is a difference in mean wait times at the different bank branches.

Note: the p-value is 0.001379.

Classical Hypothesis Testing for the Difference Between Two Dependent Population Means (unknown population standard deviation)

In this case, we are talking about dependent (or related) samples, also called *matched pairs.* This often happens when we take two measures from the same group of people: for example, giving students a pretest on a particular topic, providing them instruction, and then administering a posttest. Each student will have two values associated with him/her. In this case, we are looking to determine whether the instruction make a statistically significant difference in their test scores (presumably the scores would have increased). Consequently, we are looking at performing a t-test on the differences in the students' scores. Hence, we need to calculate each student's difference. Note that it is usually unimportant how the difference is computed (here, we could make the difference equal the DVD score minus the Book score, or vice-versa); however, the way the difference is computed should be related as one goes through the problem.

Example Which is better: Books or DVDs?

A study was done to determine whether students remember more details from a classical work of literature by reading the book or by watching a DVD movie based on a classical work of literature. Each student was given a novel to read and a DVD to watch. They were then given a 20-point quiz on each. At the 0.05 level, can we conclude that students remembered more from watching the DVD than from reading the book? Since the question asks whether the DVD scores are higher than the Book scores, we will compute the difference as Book score minus DVD score; if this is the case, then the mean difference should be positive.

Student	1	2	3	4	5	6	7
DVD	85	72	80	80	70	75	80
Book	90	80	90	75	80	90	84
Difference = Book - DVD	5	8	10	-5	10	15	4

Solution

Step 1: **State the hypotheses**

We are interested in whether students scored higher on the Book quiz than on the quiz on the DVD. This will be measured by looking at the differences between each student's Book and DVD scores: if the difference is defined as Difference = Book – DVD, we obtain the last row of the table above. The population parameter for this case is called μ_D, the mean difference. Since we expect the Book scores to exceed the DVD scores, we will hypothesize that μ_D is positive.

H_0: $\mu_D \leq 0$

H_1: $\mu_D > 0$ (this is the claim)

Step 2: **Locate the Rejection Region**

The rejection region is determined by the alternative hypothesis. Since H_1 states that $\mu_D > 0$, we have an upper tail test. In this case, we compute the mean difference, $\overline{D}$, and the standard deviation of the differences, s_D. These are computed in the same way we computed $\bar{x}$ and s earlier in the course; only the notation is slightly different.

So, $\overline{D} = \frac{\sum D}{n} = \frac{47}{7} = 6.7143$, and $s_D = \frac{\sum(D-\overline{D})^2}{n-1} = 6.317$.

A significance level of 0.10 with df = 7 – 1 = 6 in a one-tailed test corresponds to critical t-score of 1.440.

We will reject H_0 if our test value is greater than 1.440.

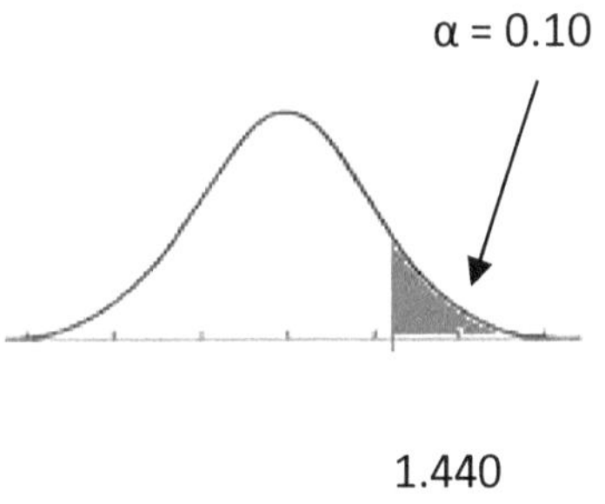

Step 3: **Calculate the Test Statistic**

We know:	Thus:
$\mu_D = 0$ $\overline{D} = 6.7143$ $s_D = 6.317$ $n = 7$ $df = n - 1 = 6$ critical $t = 1.440$	$t = \dfrac{\overline{D} - \mu_D}{s_D/\sqrt{n}} = \dfrac{6.7143 - 0}{6.317/\sqrt{7}} = \dfrac{6.7143}{2.3876} = 2.812$ 

Step 4: **Make a Decision about H_0**

Since $2.812 > 1.440$, the test statistic falls in the rejection region. We *reject* H_0.

Step 5: **State your Conclusion**

At the 0.10 level of significance, there is sufficient evidence to support the claim that the mean difference is positive; that is, that the Book scores are greater than the DVD scores.

You can see the process is the same regardless of the test statistic used. However, finding the p- value for the t-statistic is more difficult than for the z-statistic. You must use technology (the TI or *StatCrunch*) to assist. See the TI step-by-step directions for finding the p-value in this case (use the t-cdf distribution). In this case, the p-value is 0.015. Since $p < \alpha$, this confirms that we reject the null hypothesis.

Classical Hypothesis Testing for the Difference Between Two Population Proportions

Sometimes we wish to test for the difference between two population proportions (percentages). For example, we may be interested in whether the high school dropout rate the same for Fall River as New Bedford. Let's pursue this idea.

Example Dropout Rates: Fall River and New Bedford

Data from the Massachusetts Department of Education website for 2018 states that the high school dropout rate in Fall River was 14.9% (n = 686), while the New Bedford rate was 15.9% (n = 423). Although these rates appear to be different, is there a statistically significant difference between the two cities' high school dropout rates?

	Fall River	**New Bedford**
Proportion of HS dropouts	$\widehat{p}_1$ = 14.9% = .149	$\widehat{p}_2$ = 15.8% = .158
Number of Dropouts	$x_1 = \widehat{p}_1 * n_1 = 0.149 * 686 = 102$	$x_2 = \widehat{p} * n_2 = 0.158 * 423 = 67$
Sample Size (cohort size)	$n_1 = 686$	$n_2 = 423$

We need to compute a weighted percentage, called $\overline{p}$, of dropouts. We find this as follows:

$\overline{p} = \frac{x_1+x_2}{n_1+n_2} = \frac{102+67}{686+423} = \frac{169}{1109} = \mathbf{0.152}$. This looks at the total number of dropouts in both cities as a percentage of the total cohort size. And since this is really a binomial situation (either students dropped out or they didn't), we can find $\overline{q}$, the complement of the dropout percentage:

$\overline{q} = \mathbf{1} - \overline{p} = \mathbf{1} - \mathbf{0.152} = \mathbf{0.848}$. We will use both $\overline{p}$ and $\overline{q}$ in our calculations below.

Solution

Step 1: State the hypotheses

We are interested in whether there is a difference in the dropout rates in the two cities. Of course, this means we are looking at whether the proportions are the same or different. Consequently, the hypotheses are:

H_0: $p_1 = p_2$ or, $p_1 - p_2 = 0$

H_1: $p_1 \neq p_2$ or, $p_1 - p_2 \neq 0$ (this is the claim as the question asks if the rates are different)

Step 2: Locate the Rejection Region

The rejection region is determined by the alternative hypothesis. Since H_1 states that $p_1 - p_2 \neq 0$, we have a two-tailed test. As before, when dealing with proportions, we will use a z-test. Let's test at the 0.05 level.

$\hat{p}_1 = .149$ $\hat{p}_2 = .158$

$n_1 = 686$ $n_2 = 423$

$\overline{p} = 0.152$ $\overline{q} = 1 - \overline{p} = 0.848.$

$\alpha = 0.05$

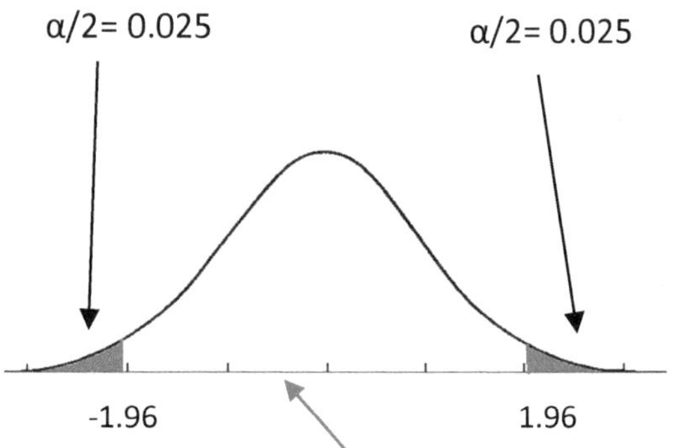

Step 3: Calculate the Test Statistic

$$z = \frac{(\hat{p}_1 - \hat{p}_2) - (p_1 - p_2)}{\sqrt{\overline{p}\overline{q}\left(\frac{1}{n_1} + \frac{1}{n_2}\right)}} = \frac{(.149 - .158) - 0}{\sqrt{(.152)(.848)\left(\frac{1}{686} + \frac{1}{423}\right)}} = \frac{-0.009}{.02219} = -0.4054.$$

Step 4: Make a Decision about H_0

Since $-1.96 < -0.4054 < 1.96$, the test statistic falls in the non-rejection region, we *fail to reject* H_0.

Step 5: State your Conclusion

At the 0.05 level of significance, there is insufficient evidence to support the claim that there is a difference in dropout rates between the two cities.

Elaine A. Previte
August 2019

Classical Hypothesis Testing for the Difference Between Two Population Variances (or Standard Deviations)

Earlier we conducted t-tests for the difference between two independent population means. We had to decide which test to use: pooled-variance or separate-variance t-tests. They were similar tests, but the degrees of freedom and standard errors were computed differently in either case, depending on whether the population variances were assumed equal or unequal. Statisticians have a way of determining whether the variances are equal or not by using a new test, the F-test, for population variances.

The F-stat is probably the easiest of all computations: $F = \frac{s_1^2}{s_2^2}$, where s_1^2 is the larger of the two variances, regardless of the subscripts of the problem's data. Note that F will *never* be negative as it is the ratio of two squares. We have two values for degrees of freedom here, dfN (numerator) and dfD (denominator): both are one less than the sample size, and both are needed to find the critical F value(s).

The F statistic is also different from the z and the t in that it is not a symmetric curve, and its shape depends on the dfN and dfD. The image below shows how different F curves appear as the number of df increase from very small (dfN = 1 and dfD = 1) to very large (dfN = 100 and dfD = 100).

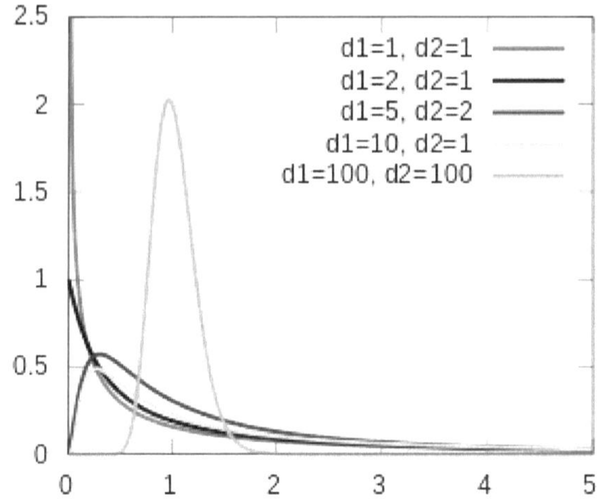

There is another important characteristic of the F statistic involves the use of hard-copy F tables. Since we need two df to find a critical F-value (see below), we need separate tables for different alpha levels (for example, α = .10 or .05 or 0.01, etc.). Be sure to check the heading of the page you're using when looking for critical F-values.

Example: Lunch-Time Lines at the Bank

As an example, let's revisit the bank branch problem. We would like to determine whether the population variances are equal or unequal. This impacts our t-test for the difference between the population means: would we use a pooled-variance t-test or a separate-variance t-test?

Recall the summary statistics:

City branch: $\bar{x}_1 = 4.28667;\ s_1^2 = 2.682995;\ n_1 = 15$
Residential branch: $\bar{x}_2 = 6.64;\ s_1^2 = 3.213447;\ n_2 = 13$

Again, recall we are testing whether the population *variances* are equal or unequal.

Step 1: **State the hypotheses**

H_0: $\sigma_1^2 = \sigma_2^2$ or $\sigma_1^2 - \sigma_2^2 = 0$

H_1: $\sigma_1^2 \neq \sigma_2^2$ or $\sigma_1^2 - \sigma_2^2 \neq 0$

Step 2: **Locate the Rejection Region**

The rejection region is determined by the alternative hypothesis. Since H_1 states that $\sigma_1^2 - \sigma_2^2 \neq 0$, we have a two-tailed test. Since this is a two-tailed test, we find the critical F-value by looking at the $\frac{\alpha}{2} = 0.025$ page of the F-table. Then we look for the degrees of freedom for the numerator. This is one less than the sample size of the sample whose variance is in the numerator of the F-statistic. In this case, dfN = 13 -1 = 12. Similarly, dfD = 15 – 1 = 14. Then we find the intersection of the proper dfN column and dfD row; this gives us the critical F-value in this situation. Here our critical F-value is 3.05. We will reject H_0 if our test value is greater than 3.05. See the partial F-table on the next page.

IV. Percentage Points of the F Distribution (continued)

$F_{0.025,\nu_1,\nu_2}$

$\alpha = 0.025$

Degrees of Freedom for the Numerator (ν_1)

Degrees of Freedom for the Denominator (ν_2)

ν_2 \ ν_1	1	2	3	4	5	6	7	8	9	10	12	15	20	24	30	40	60	120	∞
1	647.8	799.5	864.2	899.6	921.8	937.1	948.2	956.7	963.3	968.6	976.7	984.9	993.1	997.2	1001	1006	1010	1014	1018
2	38.51	39.00	39.17	39.25	39.30	39.33	39.36	39.37	39.39	39.40	39.41	39.43	39.45	39.46	39.46	39.47	39.48	39.49	39.50
3	17.44	16.04	15.44	15.10	14.88	14.73	14.62	14.54	14.47	14.42	14.34	14.25	14.17	14.12	14.08	14.04	13.99	13.95	13.90
4	12.22	10.65	9.98	9.60	9.36	9.20	9.07	8.98	8.90	8.84	8.75	8.66	8.56	8.51	8.46	8.41	8.36	8.31	8.26
5	10.01	8.43	7.76	7.39	7.15	6.98	6.85	6.76	6.68	6.62	6.52	6.43	6.33	6.28	6.23	6.18	6.12	6.07	6.02
6	8.81	7.26	6.60	6.23	5.99	5.82	5.70	5.60	5.52	5.46	5.37	5.27	5.17	5.12	5.07	5.01	4.96	4.90	4.85
7	8.07	6.54	5.89	5.52	5.29	5.12	4.99	4.90	4.82	4.76	4.67	4.57	4.47	4.42	4.36	4.31	4.25	4.20	4.14
8	7.57	6.06	5.42	5.05	4.82	4.65	4.53	4.43	4.36	4.30	4.20	4.10	4.00	3.95	3.89	3.84	3.78	3.73	3.67
9	7.21	5.71	5.08	4.72	4.48	4.32	4.20	4.10	4.03	3.96	3.87	3.77	3.67	3.61	3.56	3.51	3.45	3.39	3.33
10	6.94	5.46	4.83	4.47	4.24	4.07	3.95	3.85	3.78	3.72	3.62	3.52	3.42	3.37	3.31	3.26	3.20	3.14	3.08
11	6.72	5.26	4.63	4.28	4.04	3.88	3.76	3.66	3.59	3.53	3.43	3.33	3.23	3.17	3.12	3.06	3.00	2.94	2.88
12	6.55	5.10	4.47	4.12	3.89	3.73	3.61	3.51	3.44	3.37	3.28	3.18	3.07	3.02	2.96	2.91	2.85	2.79	2.72
13	6.41	4.97	4.35	4.00	3.77	3.60	3.48	3.39	3.31	3.25	3.15	3.05	2.95	2.89	2.84	2.78	2.72	2.66	2.60
14	6.30	4.86	4.24	3.89	3.66	3.50	3.38	3.29	3.21	3.15	3.05	2.95	2.84	2.79	2.73	2.67	2.61	2.55	2.49
15	6.20	4.77	4.15	3.80	3.58	3.41	3.29	3.20	3.12	3.06	2.96	2.86	2.76	2.70	2.64	2.59	2.52	2.46	2.40
16	6.12	4.69	4.08	3.73	3.50	3.34	3.22	3.12	3.05	2.99	2.89	2.79	2.68	2.63	2.57	2.51	2.45	2.38	2.32
17	6.04	4.62	4.01	3.66	3.44	3.28	3.16	3.06	2.98	2.92	2.82	2.72	2.62	2.56	2.50	2.44	2.38	2.32	2.25
18	5.98	4.56	3.95	3.61	3.38	3.22	3.10	3.01	2.93	2.87	2.77	2.67	2.56	2.50	2.44	2.38	2.32	2.26	2.19
19	5.92	4.51	3.90	3.56	3.33	3.17	3.05	2.96	2.88	2.82	2.72	2.62	2.51	2.45	2.39	2.33	2.27	2.20	2.13
20	5.87	4.46	3.86	3.51	3.29	3.13	3.01	2.91	2.84	2.77	2.68	2.57	2.46	2.41	2.35	2.29	2.22	2.16	2.09
21	5.83	4.42	3.82	3.48	3.25	3.09	2.97	2.87	2.80	2.73	2.64	2.53	2.42	2.37	2.31	2.25	2.18	2.11	2.04
22	5.79	4.38	3.78	3.44	3.22	3.05	2.93	2.84	2.76	2.70	2.60	2.50	2.39	2.33	2.27	2.21	2.14	2.08	2.00
23	5.75	4.35	3.75	3.41	3.18	3.02	2.90	2.81	2.73	2.67	2.57	2.47	2.36	2.30	2.24	2.18	2.11	2.04	1.97
24	5.72	4.32	3.72	3.38	3.15	2.99	2.87	2.78	2.70	2.64	2.54	2.44	2.33	2.27	2.21	2.15	2.08	2.01	1.94
25	5.69	4.29	3.69	3.35	3.13	2.97	2.85	2.75	2.68	2.61	2.51	2.41	2.30	2.24	2.18	2.12	2.05	1.98	1.91
26	5.66	4.27	3.67	3.33	3.10	2.94	2.82	2.73	2.65	2.59	2.49	2.39	2.28	2.22	2.16	2.09	2.03	1.95	1.88
27	5.63	4.24	3.65	3.31	3.08	2.92	2.80	2.71	2.63	2.57	2.47	2.36	2.25	2.19	2.13	2.07	2.00	1.93	1.85
28	5.61	4.22	3.63	3.29	3.06	2.90	2.78	2.69	2.61	2.55	2.45	2.34	2.23	2.17	2.11	2.05	1.98	1.91	1.83
29	5.59	4.20	1.61	3.27	3.04	2.88	2.76	2.67	2.59	2.53	2.43	2.32	2.21	2.15	2.09	2.03	1.96	1.89	1.81
30	5.57	4.18	3.59	3.25	3.03	2.87	2.75	2.65	2.57	2.51	2.41	2.31	2.20	2.14	2.07	2.01	1.94	1.87	1.79
40	5.42	4.05	3.46	3.13	2.90	2.74	2.62	2.53	2.45	2.39	2.29	2.18	2.07	2.01	1.94	1.88	1.80	1.72	1.64
60	5.29	3.93	3.34	3.01	2.79	2.63	2.51	2.41	2.33	2.27	2.17	2.06	1.94	1.88	1.82	1.74	1.67	1.58	1.48
120	5.15	3.80	3.23	2.89	2.67	2.52	2.39	2.30	2.22	2.16	2.05	1.94	1.82	1.76	1.69	1.61	1.53	1.43	1.31

Step 3: **Calculate the Test Statistic**

We know the following:

City branch: $s_1^2 = 2.682995;\ n_1 = 15$

Residential branch: $s_1^2 = 3.213447;\ n_2 = 13$

The F-statistic is: $F = \dfrac{larger\ variance}{smaller\ variance} = \dfrac{3.213447}{2.682995} = 1.1977,$ which does *not* fall in the rejection region. See the graph below.

Step 4: **Make a Decision about H₀**

Since 1.1977 < 3.05, the test statistic does *not* fall in the rejection region, we *fail to reject* H_0.

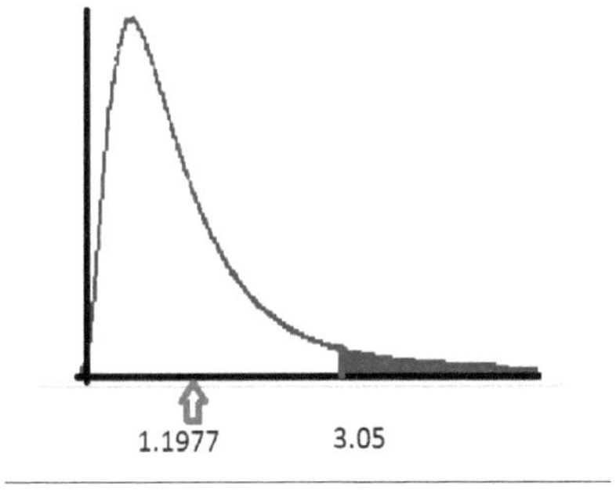

Step 5: State your Conclusion

At the 0.05 level of significance, there is insufficient evidence to reject the null hypothesis; thus, we cannot reject the hypothesis that the two population variances are equal. Consequently, we would operate under the assumption that the two population variances are equal; therefore, a pooled-variance t-test would be used to test whether the two population *means* are equal.

Elaine A. Previte
August 2019

Chi-Square Tests
Chapter 11 Summary

The Idea

Oftentimes, as we have seen, statisticians are interested in a particular hypothesis. In this case, we are interested in frequency distributions and contingency tables – do sample results meet expectations or are they significantly different from what was expected? The statistic used in these cases will be the Chi-Square (χ^2) statistic (that's "chi" like "kai," not "chai" or "key"). The symbol is the Greek letter χ, of course, and the value of χ^2 is always positive. Its distribution is asymmetrical, and critical values depend on the α level and degrees of freedom.

χ^2 Goodness-of-Fit Test

Suppose the cafeteria is allowing students to select a new lunch special. They arrange a taste-test that includes five entrées, A, B, C, D, and E. They would like to see whether or not students prefer one or more entrées. Below are the results of the taste test.

Entree	A	B	C	D	E	Total
# Students	32	28	14	16	10	100

If the students have no preference, we would expect 20 (i.e., 100 students ÷ 5 entrees) to vote for each of the entrées. Question: Are the observed values above significantly different from the number expected if there were no preference? The expected number of students has been added to the table. Let's test the hypotheses at the 0.05 level.

Entree	A	B	C	D	E	Total
# Students (Observed)	32	28	14	16	10	100
# Expected	20	20	20	20	20	100

The hypotheses are: H_0: There is no preference (claim)

H_A: At least one entrée is preferred over the others

The test statistic is χ^2, with degrees of freedom equal to number of entrées (categories) minus one, so df = 5 – 1 = 4. A portion of the χ^2 table is below, along with a graph of the χ^2 function. Note that the area of interest is in the upper tail, and that the χ^2table is similar to the t-table in that we use degrees of freedom, so consequently there are a limited number of options for critical values.

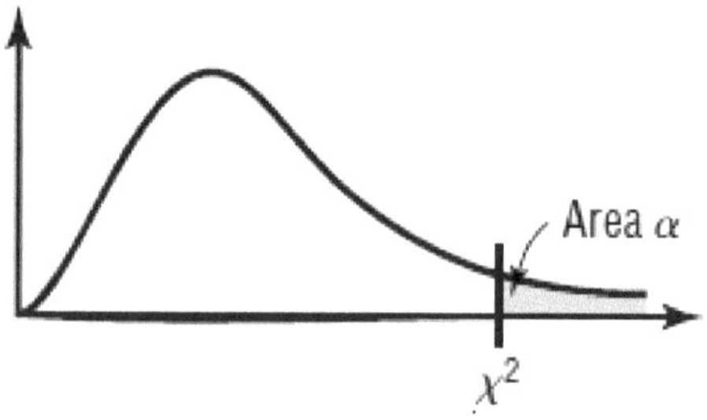

Table G The Chi-Square Distribution

Degrees of freedom	α 0.995	0.99	0.975	0.95	0.90	0.10	0.05	0.025	0.01	0.005
1	—	—	0.001	0.004	0.016	2.706	3.841	5.024	6.635	7.879
2	0.010	0.020	0.051	0.103	0.211	4.605	5.991	7.378	9.210	10.597
3	0.072	0.115	0.216	0.352	0.584	6.251	7.815	9.348	11.345	12.838
4	0.207	0.297	0.484	0.711	1.064	7.779	9.488	11.143	13.277	14.860
5	0.412	0.554	0.831	1.145	1.610	9.236	11.071	12.833	15.086	16.750
6	0.676	0.872	1.237	1.635	2.204	10.645	12.592	14.449	16.812	18.548
7	0.989	1.239	1.690	2.167	2.833	12.017	14.067	16.013	18.475	20.278
8	1.344	1.646	2.180	2.733	3.490	13.362	15.507	17.535	20.090	21.955
9	1.735	2.088	2.700	3.325	4.168	14.684	16.919	19.023	21.666	23.589
10	2.156	2.558	3.247	3.940	4.865	15.987	18.307	20.483	23.209	25.188
11	2.603	3.053	3.816	4.575	5.578	17.275	19.675	21.920	24.725	26.757

The critical value is 9.488, so if our χ^2stat exceeds that value, we will reject H_0.

To calculate our χ^2stat, we use the following formula: $\chi^2 = \sum \frac{(O-E)^2}{E}$, which means we take each observed value, subtract the expected value, square the difference, and divide by the expected value, and sum the results foe all categories. In other words,

Entree	**A**	**B**	**C**	**D**	**E**	**Total**
# Students (Observed)	32	28	14	16	10	100
# Expected	20	20	20	20	20	100
O - E	12	8	-6	-4	-10	
(O - E)²	144	64	36	16	100	
$\frac{(O-E)^2}{E}$	$\frac{144}{20}$	$\frac{64}{20}$	$\frac{36}{20}$	$\frac{16}{20}$	$\frac{100}{20}$	$\frac{360}{20} = 18$

$$\chi^2 = \sum \frac{(O-E)^2}{E} = \frac{(20-12)^2}{20} + \frac{(20-8)^2}{20} + \frac{(14-20)^2}{20} + \frac{(16-20)^2}{20} + \frac{(10-20)^2}{20} = \frac{360}{20} = 18$$

Elaine A. Previte
August 2019

Since our χ^2stat is 18 and exceeds the critical $\chi^2$of 9.488, we must reject the null hypothesis*. There is sufficient evidence to reject the claim that students have no preference of entrée. However, this test does not tell us *which* entrée(s) they *do* prefer. We will leave that for another test later.

*Remember, we reject the null hypothesis here because if there is no preference, it would be highly unlikely to observe these student results that yield a χ^2stat of 18. In fact, the p-value is 0.0012 – significantly less than our α of 0.05.

Note: the expected values do not have to be equal. They were equal in this example because our null hypothesis claimed "no preference," meaning equal preference for each entrée. Note the following example.

Example 2

A survey of BCC Business Administration Transfer majors 10 years after graduation from BCC showed that 38% were still employed by the same company they worked for upon their baccalaureate graduation; 32% had changed companies at least once; 23% were freelancing or consulting; and 7% had formed their own companies. To see if this was consistent with an earlier graduating class (who graduated 15 years ago) of Bus Admin majors, the division dean surveyed 300 graduates and found that 122 were still employed by the same company they worked for upon their baccalaureate graduation; 85 had changed companies at least once; 76 were freelancing or consulting; and 17 had formed their own companies. At α = 0.10, test the claim that the percentages are the same for the earlier class of Business Administration majors.

One thing should jump out immediately: the original data were given as percentages, and the data for the earlier graduating class were given as counts (frequencies). You cannot compare proverbial apples and oranges: you need to convert those *percentages* to *frequencies* based on the sample of 300.

	Same company	**Changed companies**	**Consulting**	**Own company**	**Total**
Observed	122	85	76	17	300
Expected	.38* 300 = 114	.32 * 300 = 96	.23 * 300 = 69	.07 * 300 = 21	300
O - E	8	-11	7	-4	
(O - E)²	64	121	49	16	
$\frac{(O-E)^2}{E}$	$\frac{64}{114}$	$\frac{121}{96}$	$\frac{49}{69}$	$\frac{16}{21}$	3.29387

We've gotten a little ahead of ourselves; let's back up. Our hypotheses are:

H_0: The students from the earlier graduating class are distributed as follows: 38% were still employed by the same company they worked for upon their baccalaureate graduation; 32% had changed companies at least once; 23% were freelancing or consulting; and 7% had formed their own companies. (claim)

H_1: The distribution is not the same as stated in the null hypothesis.

The degrees of freedom here are number of categories minus one: df = n – 1 = 4 -1 = 3; α = 0.10, so the critical value for χ^2 is 6.251.

Since $\chi^2 = 3.29387$, we fail to reject the null hypothesis. There is insufficient evidence to reject the claim that the students from the earlier graduating class are distributed as follows: 38% were still employed by the same company they worked for upon their baccalaureate graduation; 32% had changed companies at least once; 23% were freelancing or consulting; and 7% had formed their own companies.

Tests Using Contingency Tables

Recall that a contingency table is simply a table that shows two categorical (qualitative) variables. There are two types of tests that may be performed: tests for independence of variables and tests for homogeneity of proportions. In either case, the degrees of freedom are found by multiplying (the number of rows minus 1) by (the number of columns minus one); i.e., df = (#rows – 1)(#columns – 1). The χ^2 stat is computed as before: $\chi^2 = \sum \frac{(O-E)^2}{E}$; however, in this case we compute the expected value for each cell by dividing the row total by the grand total and then multiplying by the column total. By dividing the row total by the grand total, we get the percentage of the total sample that is included in that particular row. By then multiplying by the column total for each major, we get the expected number of participants in that row who should be in that column if there were no relationship between the two variables. In other words, we're simply allocating numbers of elements in each row to each column on a proportional basis.

χ^2 Test for Independence

Recall that independence means that one variable is not affected by another variable. For example, one might be interested in whether gender affects choice of college major. Suppose a sample of 100 BCC students is taken, and students are asked to disclose their gender and major program. These data are listed in the contingency table below. At the 0.05 level, do the data show that choice of major is independent of gender?

	Accounting	Marketing	Finance	Human Resources	Total
Female	10	22	5	23	60
Male	20	8	10	2	40
Total	30	30	15	25	100

H_0: Choice of academic major is independent of gender (claim)

H_1: There is no relationship between gender and choice of major

df = (#rows – 1)(#columns – 1) = (2 – 1) (4 – 1) = (1)(3) = 3

$\alpha = 0.05$

critical $\chi^2 = 7.815$

The expected number of female accounting majors: $E = \left(\frac{60}{100}\right) \cdot 30 = 18$. In other words, since 60% of the sample is female, if there is no relationship between gender and major as hypothesized, we would expect that 60% of the accounting majors should be female.

The expected number of female marketing majors: $E = \left(\frac{60}{100}\right) \cdot 30 = 18.$

The expected number of female finance majors: $E = \left(\frac{60}{100}\right) \cdot 15 = 9.$

The expected number of female human resource majors: $E = \left(\frac{60}{100}\right) \cdot 25 = 15.$

Since males represent 40% of the sample, we would expect 12 accounting majors, 12 marketing majors, 6 finance majors, and 10 human resource majors. So,

$$\chi^2 = \sum \frac{(O-E)^2}{E} = \frac{(10-18)^2}{18} + \frac{(22-18)^2}{18} + \frac{(5-9)^2}{9} + \frac{(23-15)^2}{15} + \frac{(20-12)^2}{12} + \frac{(8-12)^2}{12} + \frac{(10-6)^2}{6} + \frac{(2-10)^2}{10} = 26.222$$

Since our χ^2 > critical χ^2, we reject H_0 (i.e. 26.222 > 7.815; note the p-value is 0.00000856, which is far less than our α of 0.05). There is sufficient evidence to reject the claim that gender and choice of major are independent.

χ^2 Test for Homogeneity of Proportions

Another χ^2 test involving contingency tables is called a test for homogeneity of proportions; i.e. are the proportions from our samples equal or is at least one of them different? The null hypothesis is H_0: $p_1=p_2= \ldots = p_n$. The alternative hypothesis is H_1: at least one of the proportions is different. The degrees of freedom are found as above: df = (#rows – 1)(#columns – 1).

Example

A consumer group was interested in finding out if three parcel delivery services lose packages at the same rate. They sample 100 customers from each service; the results are below.

	Service 1	Service 2	Service 3	Total
Yes	10	7	4	21
No	90	93	96	279
Total	100	100	100	**300**

H_0: $p_1=p_2= p_3$

H_1: at least one of the proportions is different

df = (#rows – 1)(#columns – 1) = (2 – 1) (3 – 1) = (1)(2) = 2

$\alpha = 0.05$

critical $\chi^2 = 5.991$

Since $\frac{21}{300} = 0.07 = 7\%$ of customers in the sample lost their packages, we're testing whether any of the delivery services has a lost-package rate that is significantly different than 7%. The expected number of customers who lost packages is 7 per service (7% of 100), and the expected number of customers who didn't lose packages is 93 per delivery service. Hence,

$$\chi^2 = \sum \frac{(O-E)^2}{E} = \frac{(10-7)^2}{7} + \frac{(7-7)^2}{7} + \frac{(4-7)^2}{7} + \frac{(90-93)^2}{93} + \frac{(93-93)^2}{93} + \frac{(96-93)^2}{93}$$
$$= 2.765$$

Since our χ^2 < critical χ^2, we fail to reject H_0 (i.e. $2.765 < 5.991$; note the p-value is 0.25095, which is far greater than our α of 0.05). There is insufficient evidence to reject the claim that each delivery service's lost-package rate is equal.

Elaine A. Previte
August 2019

Analysis of Variance
Chapter 12 Summary

The Idea

In earlier chapters, we performed hypothesis tests comparing two means. But sometimes we need to examine the relationships between more than two means. For example, a pharmaceutical company may be testing several versions of a new cancer treatment in order to determine whether they're equally effective or if some of the treatments are more effective than the others. In other words, do the different treatments all come from the same population, or is/are some of them from a different population?

Statisticians don't use pair-wise hypothesis tests when looking at more than two means for several reasons. First, if you compare two means at a time, you are ignoring the other means. Second, the alpha-level may compound beyond that which is acceptable. Third, the more means you have, the more pair-wise tests you need to complete (3 means yield 3 tests; 5 means yield 10 tests; 10 means yield 45 tests!). Consequently, statisticians have developed something called analysis of variance (ANOVA for short) to help us perform these analyses. We will look at one-way ANOVA, although there are other forms (e.g., two-way ANOVA). We will use the F test to perform our analyses.

Assumptions for ANOVA F tests

1. The populations from which the samples were obtained must be normally or approximately normally distributed.
2. The samples must be independent of one another.
3. The variances of the populations must be equal.

The F test looks at a comparison of variances, even though the hypotheses deal with the relationship between several means. The null hypothesis is H_0: $\mu_1 = \mu_2 = \mu_3 = \ldots = \mu_n$, and states that all the data come from the same population. The alternative hypothesis is H_1: at least one mean is different from the others, meaning not all of the data come from the same population.

Note: if you reject H_0, you do not learn *which* mean(s) is/are different; you simply learn that not all means are equal. There are additional tests that can help you determine which mean(s) is/are different. We'll look at some of those later.

Again, the F test compares two variances, but what are they? The first is called the "between group" variance – it measures how each of the individual means vary from the overall (the so-called "grand") mean (see below). This helps determine if the data come from the same population or not.

Possible Relationships Between Individual Means and The Overall (Grand) Mean		
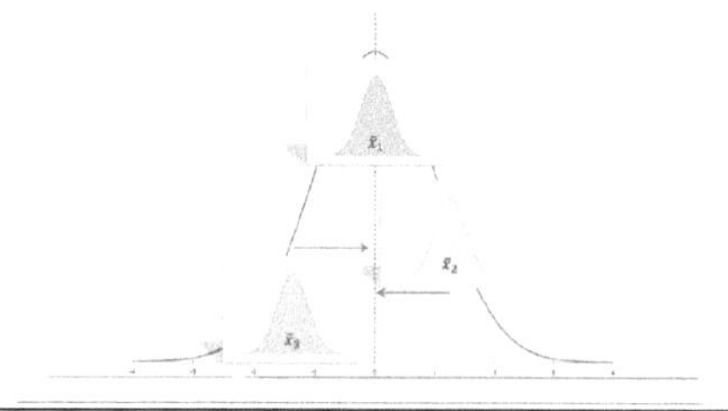	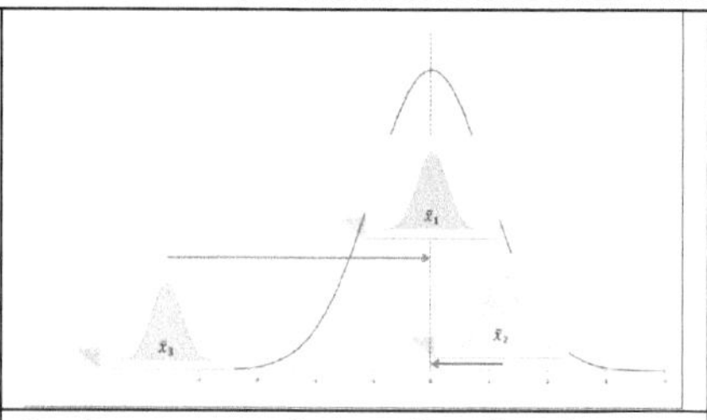	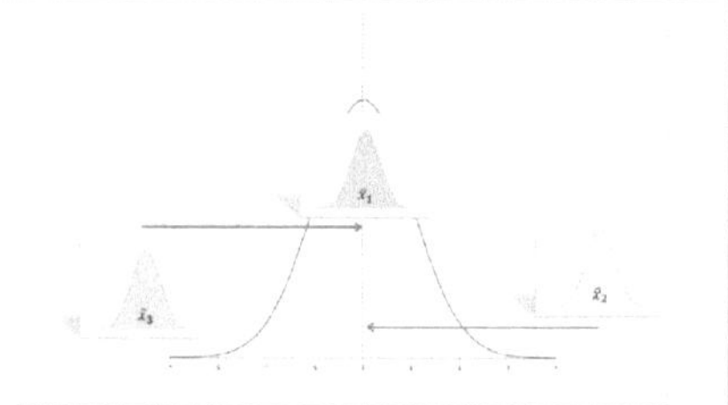
Are the means of the individual distributions "close to" the mean of the combined distribution?	Is the mean of one distribution "far away" from the mean of the combined distribution?	Are all the means of the individual distributions "far away" from the mean of the combined distribution?

Graphics from B. Foltz, Statistics 101: ANOVA, A visual Introduction https://www.youtube.com/watch?v=0Vj2V2qRU10

The other variance is called the "within-group" variance. This is a weighted average of all the sample variances and is not affected by the difference in the means. If these two ratios are approximately equal, the F stat will be approximately equal to one and the null hypothesis will not be rejected. If the two ratios are very different, then the between-group variance will be significantly larger than the within-group variance, and the F stat will be significantly larger than one and the null hypothesis will be rejected.

ANOVA Example

A study was done to compare the cost to drive 25 miles for different types of hybrid vehicles. The cost of a gallon of gas at the time of the study was approximately S2.50. Based on the information given below for different models of hybrid cars, trucks, and SUVs, is there sufficient evidence to conclude a difference in the mean cost to drive 25 miles? Use $\alpha = 0.05$. Note: dfN = k-1= 3 – 1 = 2 and dfD = N – k = 12 – 3 = 9.

H_0: $\mu_1 = \mu_2 = \mu_3$

H_1: at least one mean is different (claim) Critical F = 4.256

Step 1: Find the mean and standard deviation of each sample. Note: k = #categories = 3.

Hybrid Cars	Hybrid SUVs	Hybrid Trucks
2.10	2.10	3.62
2.70	2.42	3.43
1.67	2.25	
1.67	2.10	
1.30	2.25	
$\bar{x}_1 = 1.888$	$\bar{x}_2 = 2.224$	$\bar{x}_3 = 3.525$
$s_1 = 0.5350$	$s_2 = 0.1328$	$s_3 = 0.1344$
$n_1 = 5$	$n_2 = 5$	$n_3 = 2$

Step 2 **Find the Grand Mean, $\bar{X}_{GM}$ (the mean of all the data). Note N = $n_1 + n_2 + n_3$.**

$$\bar{X}_{GM} = \frac{2.10 + 2.70 + 1.67 + 1.67 + 1.30 + 2.10 + 2.42 + 2.25 + 2.10 + 2.25 + 3.62 + 3.43}{12}$$

$\bar{X}_{GM}$ =2.3008

Step 3 **Find the between-group variance, s_B^2.**

$$s_B^2 = \frac{\sum n_i(\bar{x}_i - \bar{X}_{GM})^2}{k-1} = \frac{n_1(\bar{x}_1 - \bar{X}_{GM})^2 + n_2(\bar{x}_2 - \bar{X}_{GM})^2 + n_3(\bar{x}_3 - \bar{X}_{GM})^2}{3-1}$$

$$= \frac{5(1.888 - 2.30)^2 + 5(2.224 - 2.30)^2 + 2(3.525 - 2.30)^2}{3-1} = 1.93942.$$

Step 4 **Find the within-group variance, s_W^2.**

$$s_W^2 = \frac{\sum(n_i - 1)s_i^2}{\sum(n_1 - 1)} = \frac{(n_1 - 1)s_1^2 + (n_2 - 1)s_2^2 + (n_3 - 1)s_3^2}{(n_1 - 1) + (n_2 - 1) + (n_3 - 1)}$$

$$= \frac{(5-1)(0.5350)^2 + (5-1)(0.1328)^2 + (2-1)(0.1344)^2}{(5-1)+(5-1)+(2-1)}$$

$$= 0.13707$$

Step 5 **Compute the F statistic. Note: dfN = k-1= 3 – 1 = 2 and dfD = N – k = 12 – 3 = 9.**

$$F = \frac{s_B^2}{s_W^2} = \frac{1.93942}{0.13707} = 14.149$$

Since the F statistic is greater than the critical F value, we reject the null hypothesis. There is sufficient evidence to support the claim that there is a difference in the mean cost of operating the different hybrid models.

Again, note that we cannot state which mean(s) is(are) different; we can say only that there is a statistically significant difference in the mean cost of operating these vehicles.

ANOVA Tables

We can do our ANOVA computations in a number of ways: Excel, StatCrunch, TI-84, etc. After one enters the above values into a TI-84+ (Stat → Edit → List 1 for the first list, List 2 for the second, and List 3 for the third), we can then perform the ANOVA test (Stat → Tests → ANOVA). Remember to enter the names of the lists in which you entered the data. So, assuming you entered the above data in L1, L2, L3, you should enter ANOVA (L1, L2, L3). Then hit enter. You should see the following results (your screen may be slightly different, depending on the model calculator you have or program you use):

One-way ANOVA

F = 14.14889758
p = .0016653943

s_{xp} = .3702326596

	Factor *(between-group)*	**Error** *(within-group)*
df	2	9
SS	3.878841667	1.23365
MS	1.939420833	0.1370722222

This table gives us the calculated F-statistic (14.14889758) as we obtained above, the p-value (0.0016653943), the between-group variance ($s_B^2 = 1.939420833$), and the within-group variance ($s_W^2 = 0.1370722222$). This saves a lot of hand calculations!

Optional: Comparing Multiple Pairs of Means = 0.05

In the case where we reject H_0 in an ANOVA test, we conclude there is a difference in the means. However, ANOVA itself *doesn't* tell us which mean(s) are different from the rest. Consequently, statisticians have created some tests where we can compare pairs of means to determine which one(s) are different from the rest. We will examine the Tukey-Kramer Multiple Comparison Test for One-Way ANOVA, using α = 0.05.

The three sets of hypotheses are:

H_0: $\mu_1 = \mu_2$
H_1: $\mu_1 \neq \mu_2$

H_0: $\mu_1 = \mu_3$
H_1: $\mu_1 \neq \mu_3$

H_0: $\mu_2 = \mu_3$
H_1: $\mu_2 \neq \mu_3$

First, we need to compute the absolute sample mean differences, $|\bar{x}_1 - \bar{x}_2|$, $|\bar{x}_1 - \bar{x}_3|$ and $|\bar{x}_2 - \bar{x}_3|$.

$|\bar{x}_1 - \bar{x}_2| = |1.888 - 2.224| = 0.336$

$|\bar{x}_1 - \bar{x}_3| = |1.888 - 3.525| = 1.637$

$|\bar{x}_2 - \bar{x}_3| = |2.224 - 3.525| = 1.301$

Elaine A. Previte
August 2019

Next, we calculate the Tukey-Kramer Critical Range.

$CR_{i,j} = Q_\alpha\sqrt{\frac{MSW}{2}\left(\frac{1}{n_i}+\frac{1}{n_j}\right)}$ where $CR_{i,j}$ = The critical range for comparing sample *i* to sample *j*

Q_α = The critical value from the studentized range table below

MSW = The within-group mean square

n_i = the sample size from sample *i*

n_j = the sample size from sample *j*

The critical value for Q_α can be found using two degrees of freedom: df1 = k (number of samples) and df2 = N – k = Total number of pieces of data – number of samples. In this case, k = 3 and N = 5 + 5 + 2 = 12, so df1 = 3, df2 = 12 – 3 = 9, and $Q_\alpha = 3.88$.

The Tukey-Kramer values are:

$$CR_{1,2} = 3.88\sqrt{\frac{.1370722222}{2}\left(\frac{1}{5}+\frac{1}{5}\right)} = 0.1655730789$$ (Note: $n_1 = n_2 = 5$)

$$CR_{1,3} = CR_{2,3} = 3.88\sqrt{\frac{.1370722222}{2}\left(\frac{1}{5}+\frac{1}{2}\right)} = 0.2190325952$$ (Note: $n_1 = n_2 = 5$ and $n_3 = 2$)

Perform the hypothesis tests:

We will reject a null hypothesis if the absolute mean differences are greater than the Tukey-Kramer values.

$H_0: \mu_1 = \mu_2$ $H_1: \mu_1 \neq \mu_2$	$H_0: \mu_1 = \mu_3$ $H_1: \mu_1 \neq \mu_3$	$H_0: \mu_2 = \mu_3$ $H_1: \mu_2 \neq \mu_3$
$\lvert\bar{x}_1 - \bar{x}_2\rvert$ = 0.336 > 0.1656	$\lvert\bar{x}_1 - \bar{x}_3\rvert$ = 1.637 > 0.2190	$\lvert\bar{x}_2 - \bar{x}_3\rvert$ = 1.301 > 0.2190
Reject H_0 Conclude $\mu_1 \neq \mu_2$	Reject H_0 Conclude $\mu_1 \neq \mu_3$	Reject H_0 Conclude $\mu_2 \neq \mu_3$

Since we reject all the null hypotheses, we conclude that the three samples come from different populations.

Elaine A. Previte
August 2019

Critical values of the studentized range for $\alpha = 0.05$

df1

df2

Row	var1	var2	var3	var4	var5	var6	var7	var8	var9	var10
3	2	6.09	8.33	9.8	10.88	11.74	12.44	13.03	13.54	13.99
4	3	4.5	5.91	6.82	7.5	8.04	8.48	8.85	9.18	9.46
5	4	3.93	5.04	5.76	6.29	6.71	7.05	7.35	7.6	7.83
6	5	3.64	4.6	5.22	5.67	6.03	6.33	6.58	6.8	6.99
7	6	3.46	4.34	4.9	5.3	5.63	5.9	6.12	6.32	6.49
8	7	3.34	4.16	4.68	5.06	5.36	5.61	5.82	6	6.16
9	8	3.26	4.04	4.53	4.89	5.17	5.4	5.6	5.77	5.92
10	9	3.2	3.95	4.41	4.76	5.02	5.24	5.43	5.59	5.74
11	10	3.15	3.88	4.33	4.65	4.91	5.12	5.3	5.46	5.6
12	11	3.11	3.82	4.26	4.57	4.82	5.03	5.2	5.35	5.49
13	12	3.08	3.77	4.2	4.51	4.75	4.95	5.12	5.27	5.39
14	13	3.06	3.73	4.15	4.45	4.69	4.88	5.05	5.19	5.32
15	14	3.03	3.7	4.11	4.41	4.64	4.83	4.99	5.13	5.25
16	15	3.01	3.67	4.08	4.37	4.59	4.78	4.94	5.08	5.2
17	16	3	3.65	4.05	4.33	4.56	4.74	4.9	5.03	5.15
18	17	2.98	3.63	4.02	4.3	4.52	4.7	4.86	4.99	5.11
19	18	2.97	3.61	4	4.28	4.49	4.67	4.82	4.96	5.07
20	19	2.96	3.59	3.98	4.25	4.47	4.65	4.79	4.92	5.04
21	20	2.95	3.58	3.96	4.23	4.44	4.62	4.77	4.9	5.01
22	24	2.92	3.53	3.9	4.17	4.37	4.54	4.68	4.81	4.92
23	30	2.89	3.49	3.85	4.1	4.3	4.46	4.6	4.72	4.82
24	40	2.86	3.44	3.79	4.04	4.23	4.36	4.52	4.63	4.73
25	60	2.83	3.4	3.74	3.98	4.16	4.31	4.44	4.55	4.65
26	120	2.8	3.36	3.68	3.92	4.1	4.24	4.36	4.47	4.56
27	infinity	2.77	3.31	3.63	3.86	4.03	4.17	4.29	4.39	4.47

Elaine A. Previte
August 2019

Linear Regression and Correlation
Chapter 13 Summary

The Idea

Linear regression is the process of fitting a line to a scatterplot of points (ordered pairs) of bivariate data (x and y). For example, the scatterplot below shows apartment size (in square feet) vs. rent ($). The data for each apartment is an ordered pair of the form (area, rent). As you can see, apartments that have smaller square footage seem to also have lower rent, whereas larger apartments tend to have higher rents. Linear regression will find a liner equation that captures, as best as possible, the relationship between x and y. Thus, when x and y have a linear relationship, their relationship will be described by the equation $y = mx + b$. Linear regression computes the values of m (the slope) and b (the y-intercept) of the equation. The equation can then be used to predict rent amounts for given apartment sizes. **Correlation**, on the other hand, measures the strength of that linear relationship.

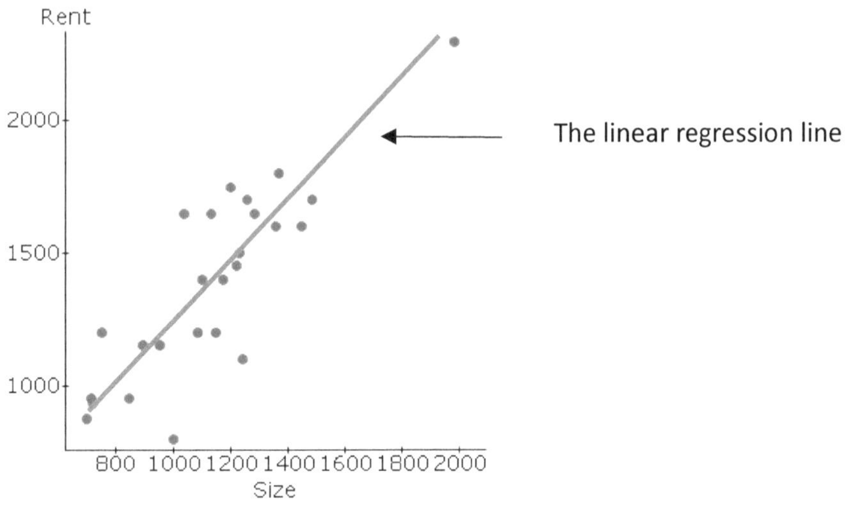

Correlation

Before computing the linear regression equation, we need to get an idea of how the two variables in question are related, if they are related at all. This is called the correlation of x and y.

Consider the two graphs on the next page – they are quite different!

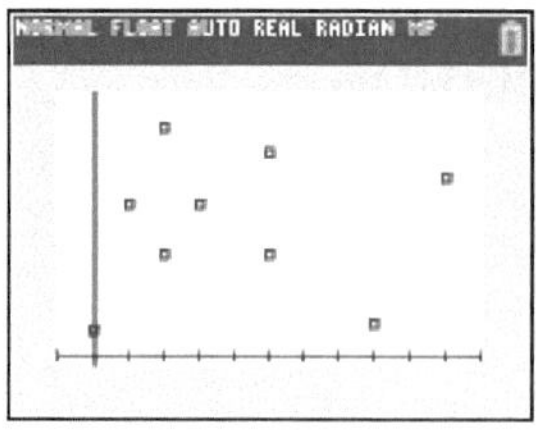

Graph A

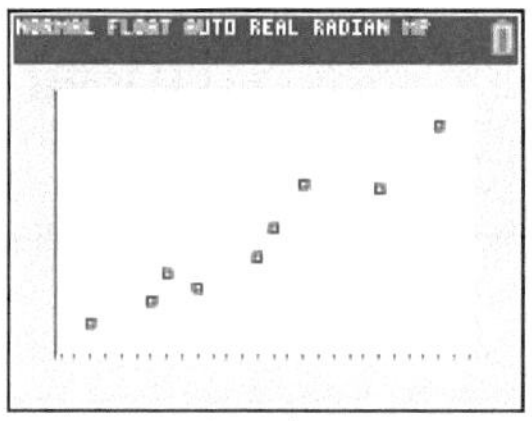

Graph B

In Graph A, regardless of the value of x, y can take on many different values. We can see there is not an apparent pattern to the bivariate data (the x-values and y-values). Therefore, the value of y is unrelated to the value of x; we say that y is *not correlated* with x.

However, in Graph B, lower values of x are associated with lower values of y and higher values of x are associated with higher values of y (this is a scatterplot of the new home square footage vs cost data below). We can see here that there is a noticeable linear pattern to this bivariate data, although this pattern is not perfectly linear. We say that x and y are correlated. (For further information on correlation, see Chapter 3.)

Definition of correlation

- The correlation between x and y is a measure of the strength of the linear relationship between x and y.
- The correlation coefficient, *r*, is a numerical measure of the correlation between x and y, and must be a number between -1 and +1, inclusive. The closer r is to -1 or +1, the stronger the relationship between x and y. The closer *r* is to 0, the weaker the relationship.
- The formula for the correlation coefficient is:

$$r = \frac{\sum xy - n\bar{x}\bar{y}}{\sqrt{(\sum x^2 - n\bar{x}^2)(\sum y^2 - n\bar{y}^2)}}$$ (see alternative formulas on page 9 of this chapter)

Note: there is a slightly different version of this formula in Chapter 3; however, they are algebraically equivalent and will yield the same value for *r*.

In order to illustrate the concept of correlation, let us examine some data. Below is sample data on the square footage of new homes (in hundreds of square feet) and their prices (in thousands of dollars):

x, sq. footage	26	27	41	29	33	34	36	45	22
y, price	235	273	387	253	295	335	395	475	203

To calculate the correlation coefficient, we'll use the last version of the formula for *r*. Using your TI-84+ calculator:

- Use STAT → EDIT to input the x-values into L1 and the y-values into L2.
- Now calculate two-variable statistics by using STAT → CALC → 2-var stats.

Now you will see that everything you need to compute *r* is on your screen.

Verify for yourself that $\sum xy = 97871$, $n = 9$, $\bar{x} = 32.556$, $\bar{y} = 316.778$, $\sum x^2 = 9977$, and $\sum y^2 = 965{,}641$. Let's input these values into the formula for *r*.

$$r = \frac{\sum xy - n\bar{x}\bar{y}}{\sqrt{(\sum x^2 - n\bar{x}^2)(\sum y^2 - n\bar{y}^2)}} = \frac{97871 - (9)(32.556)(316.778)}{\sqrt{(9977 - (9)(32.556)^2)(965{,}641 - (9)(316.778)^2)}}$$

$$r = \frac{5053.779}{5232.147} = 0.966.$$

Notice that r = 0.966 is very close to one, so we would say that we have a very strong (nearly perfect) positive linear correlation. This suggests that the cost of these new homes is strongly related to their square footage. You can see this by looking at the graph below. The line of best fit is included to illustrate the strength of this linear relationship.

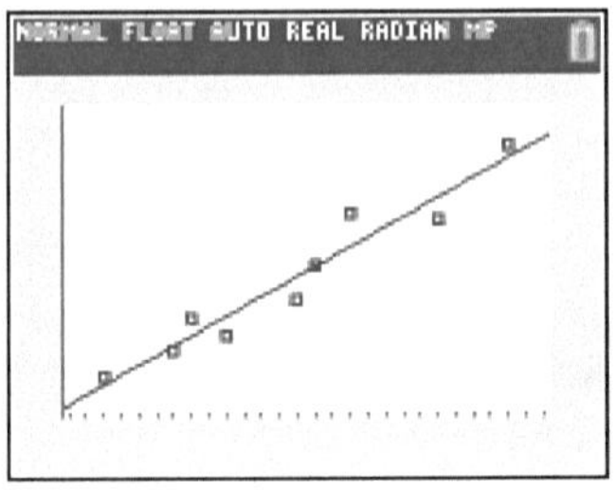

Square footage of new homes vs. construction cost

We can get an idea of the shape of the data when we know the correlation coefficient. Some examples are pictured below.

	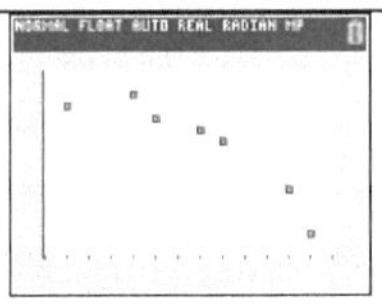	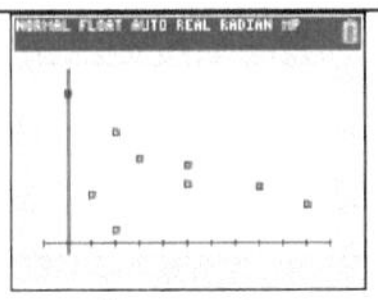	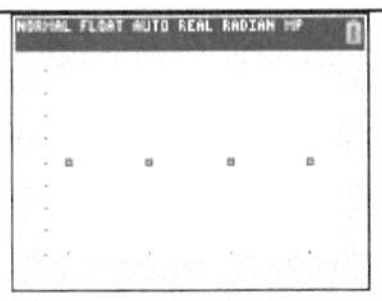	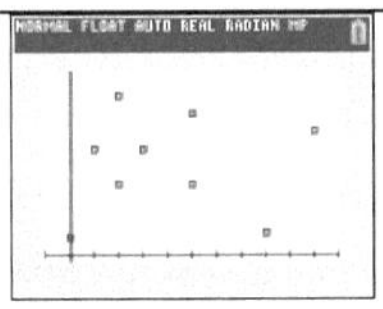	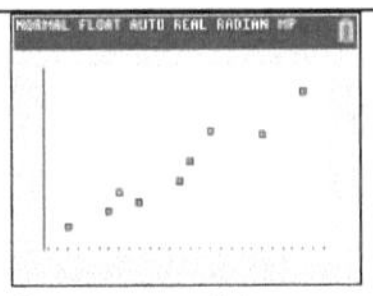	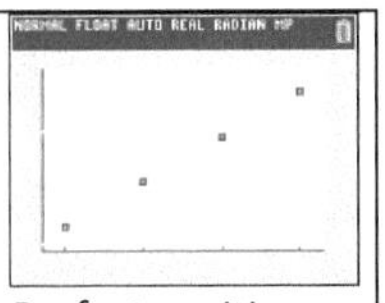
ct negative *correlation;*	*Strong negative linear correlation; r = -0.9085*	*Weak negative linear correlation; r = -0.4434*	*No correlation; r = 0*	*Weak positive linear correlation; r = 0.067*	*Strong positive linear correlation; r = 0.9659*	*Perfect positive linear correlation; r = +1*

Simple Linear Regression

Whereas the correlation coefficient will tell the strength of the linear relationship between x and y, linear regression will quantify and describe this relationship by defining a linear equation of the form $y = mx + b$. It is important to note that, in this form, y is assumed to be influenced by x; you may recall from basic algebra the phrase, "y is dependent on x." We call y the dependent variable and x is the independent variable. As in the example above, the cost of a new home is dependent on its square footage, not the other way around.

The values that must be calculated in a linear regression are *m* (the slope) and *b* (the y-intercept). We will calculate these values for the home vs. cost data given above. But first we need to further explain the idea behind those calculations.

The linear regression line is also called the "line of best fit." This name is appropriate because the regression calculations are such that they minimize the squared distances of the actual points from the values predicted by the regression equation.

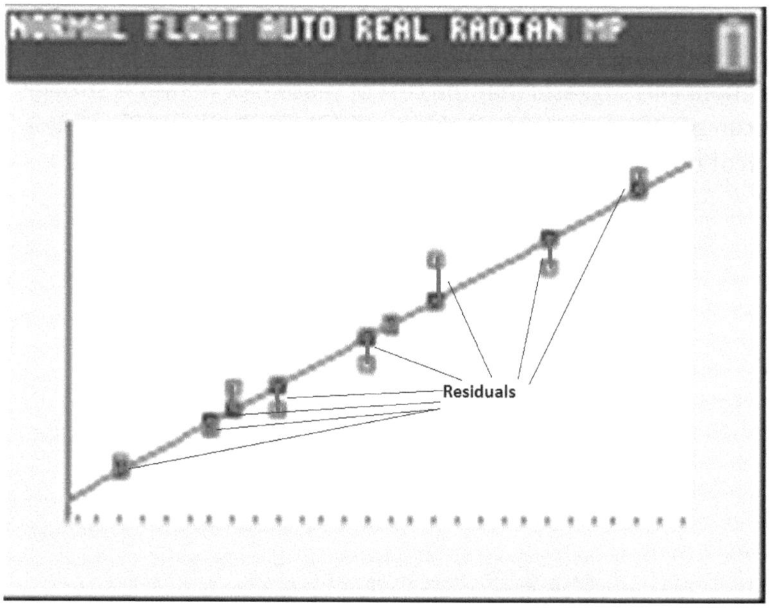

The graph above depicts the actual data points (off the line) and the predicted y-values on the regression line. The difference between the actual y-value and the predicted y-value is called the *residual* (depicted by the short vertical lines between the actual points and the predicted points). The TI screenshot below shows the actual x-values (L1), actual y-values (L2), predicted y-values (L3), and the residual values (the differences between the actual y-values and the predicted y-values) in L4. It is important to note that the sum of the residuals is (practically) zero, as shown by the screenshot on the right of the lists.

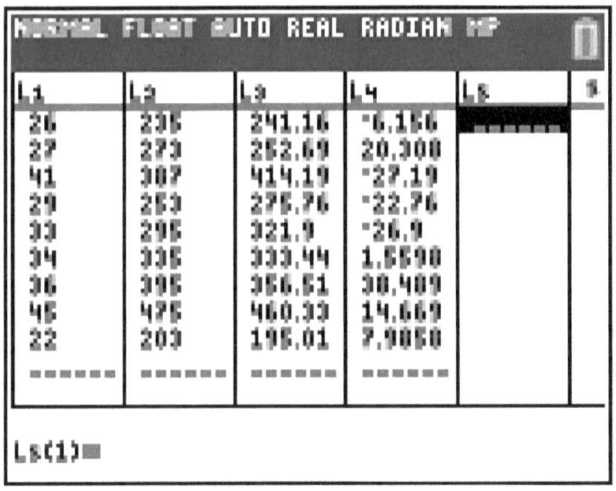

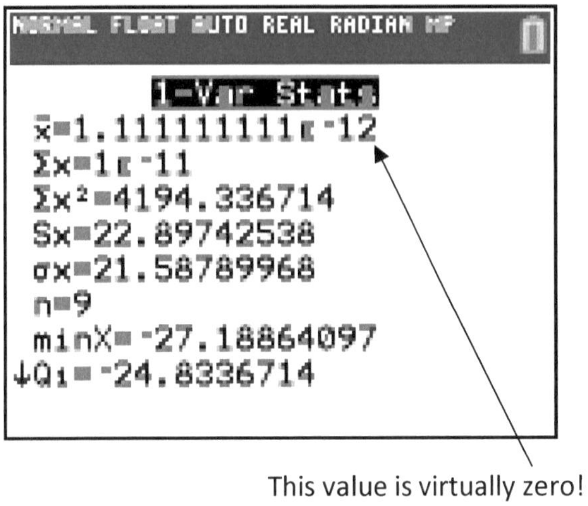

Linear regression calculates for *m* and *b* in such a way that the sum of the squared residuals is a minimum. The formulas for m and b are derived using methods from Calculus, but their formulas are:

$$m = \frac{\sum xy - n\bar{x}\bar{y}}{\sum x^2 - n\bar{x}^2} \quad \text{and} \quad b = \bar{y} - m\bar{x}.$$

Let's try this on our home vs. cost data. We know from our calculation for *r* above that $\sum xy = 97{,}871,\quad n = 9,\quad \bar{x} = 32.556,\quad \bar{y} = 316.778,\quad \sum x^2 = 9977,$ and $\sum y^2 = 965{,}641.$

Now we can use this information to calculate *m* first:

$$m = \frac{\sum xy - n\bar{x}\bar{y}}{\sum x^2 - n\bar{x}^2} = \frac{97{,}871 - 9(32.556)(316.778)}{9977 - 9(32.556)^2} = \frac{5053.778}{437.962} = 11.539.$$

This is the slope of our regression line. We use this value in our calculation of the y-intercept, *b*.

$$b = \bar{y} - m\bar{x} = 316.778 - 11.539(32.556) = -58.886.$$

Thus, our regression equation is: $\hat{y} = 11.539x - 58.886$. We put a "hat" over the *y* in the regression equation to indicate that it is the predicted y-value, not the actual y-value of the original data.

Using the Regression Line

- It is important to note that the slope of the regression line has meaning that relates to the context of our problem. The slope is the amount that y is predicted to increase when x increases by 1 unit. In our example, x is measured in hundreds of square feet and y is measured in thousands of dollars. So, when x increases by 1 unit (100 square feet), y is predicted to increase by 11.539 units (thousands of dollars).

- The meaning of the y-intercept is that it is the predicted value of y when x equals zero. Many times (like this time) the y-intercept doesn't make sense (who would build a new home with zero square feet??).

- We use the regression equation to predict y-values for given x-values. In this case, we can predict the cost of a home with particular square footage. For example, to predict the cost of a home with 40 (that is 40-hundred square feet, otherwise known as 4,000 square feet), we simply substitute 40 for x in our regression equation:

  $\hat{y} = 11.539(40) - 58.886 = 402.674$, or \$402,674. So, a house with 4,000 square feet of living space is predicted to cost \$402,674 using this model.

- The TI-84 calculator can calculate the regression equation. After inputting the data into L1 and L2, using Stat → Edit, we use Stat → Calc → 4:LinReg (ax + b).

- You can also graph the regression line by storing it in one of the equations found in the Y = graphing screen by doing the following: Y = → y_1 → VARS → 5:Statistics → EQ → 1: RegEQ. This will store the regression equation in y_1. When you plot the data using 2nd → StatPlot and choosing the Scatterplot option, you will see the actual data as well as the regression line. Note: you may also store the regression equation while you calculate it (see bullet above) by simply telling the TI to store the regression equation in y_1, for example, by using the steps as outlined above.

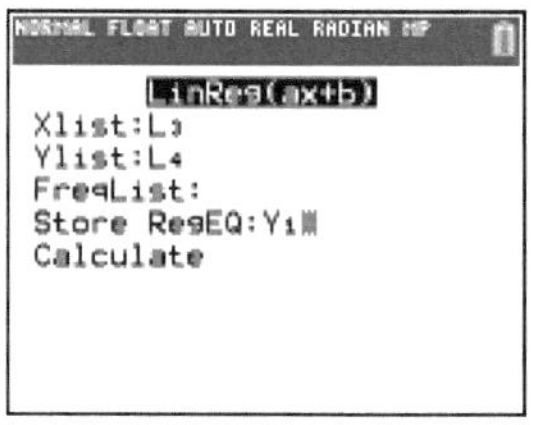

TI 84+ CE screenshot

Coefficient of Determination

Once the regression equation is calculated, we want to get an idea of how well (or how badly) the regression line fits the data. In other words, bivariate data that falls nearly on a straight line could be expected to be predicted reliably, from the regression equation; whereas data that don't follow a linear pattern could not be reliably predicted by the linear regression equation. Thus, we would like to find a number that could represent the percent of data accounted for by the regression line. We call this number the *coefficient of determination*.

Because we have our regression line, the total sum of squares of our dependent variable can now be split up into the sum of squares due to the regression plus the sum of squares due to residuals (the random part, sometimes called the error). Thus, $SS_y = SS_{\hat{y}} + SS_{residual}$.

If we divide both sides of this equation by SSy, we get

$$SS_y = SS_{\hat{y}} + SS_{residual}$$

$$\frac{SS_y}{SS_y} = \frac{SS_{\hat{y}}}{SS_y} + \frac{SS_{residual}}{SS_y}$$

$$1 = \frac{SS_{\hat{y}}}{SS_y} + \frac{SS_{residual}}{SS_y}$$

Or, we could say that, in percent terms,

100% of the variation in y	=	the percent of the variation due to the regression	+	the percent due to error

The first part of the right-hand side of the equation is called the *coefficient of determination.* That larger this value is (and it cannot be greater than 100% or 1.00), the better is the linear regression fit to the data. The smaller it is (and it cannot be less than 0% or 0), the worse the fit.

Going back to the (sq. ft, cost) data, by making a simple modification on the TI-84+, we can calculate the coefficient of determination.

First, hit 2nd catalog →DiagnosticOn.

Second, recalculate the regression.

Now notice that two additional numbers appear. One is "*r.*" This is simply the correlation coefficient that we calculated above (that's right – we won't have to calculate it anymore). But the second value is called "r^2," the coefficient of determination. It gives the percent of the y-variable that can be "explained" by the regression line. A value over .95, or 95% is a very good fit. In our example, $r = 0.9659$ and $r^2 = 0.932899$.

The coefficient of determination is nicknamed "*r-square*", or "r^2" for a good reason. The coefficient of determination is literally the square of the correlation coefficient. Try is with the values that you calculated, and you'll see that:

Coefficient of determination = (correlation coefficient)^2, or r-square = (r)^2.

$$r^2 = (r)^2 = (0.9659)^2 = 0.93296$$

(Note: the slight discrepancy between this value and the r^2 above is due to rounding the r-value before squaring it).

Since our regression explains 93.29% of the variation, 6.71% of the variation is not explained and is due to random error.

It is also important to note that the correlation coefficient between x and y can be negative, depending on whether the correlation is negative; however, the r-square must lie between 0 and 1. The two values, r and r^2, are related, but have different meanings. Be mindful of this.

Prediction

Following from the third bullet above, one of the primary uses of the regression equation is to make predictions of y-values for given values of x. Using our housing example, suppose we want to predict the cost of a home with 3500 square feet (i.e., x = 35). We would simply use our regression equation and input 35 for the value of x:

$$\hat{y} = 11.539x - 58.886$$

or, stated in context, predicted cost of house = 11.539 * square footage - 58.886

predicted cost of house = 11.539 * 35 - 58.886

= 344.979 (i.e., $349,979).

Thus, our regression equation predicts that a 3500-square foot home would cost approximately $349,979. A builder could use this information to provide customers with the approximate cost of a home.

It is important to note that predictions will be most reliable within the range of the x-values used to calculate the regression equation. Since our x-values ranged from 22 to 45 (i.e., 2200 to 4500 square feet), the predicted y-values would be more reliable for x-values in this range. It is still possible to do predict y-values for x-values outside this range (for example, we could predict the cost of a house that would have 10,000 square feet), but they would not be expected to be reliable from a statistical point of view.

Additional Information

Equivalent forms of Formulae

Since the formulae presented above contain summation notation, we can use algebraic techniques to rewrite those formulae in equivalent forms. For example, an equivalent formula for the correlation coefficient is as follows:

$$r = \frac{COV(x,y)}{s_x \cdot s_y} = \frac{\frac{\sum(x-\bar{x})(y-\bar{y})}{n-1}}{s_x \cdot s_y}.$$

An equivalent form of the regression equation's slope formula is: $m = \frac{\sum[(x-\bar{x})(y-\bar{y})]}{\sum(x-\bar{x})^2}$.

Hypothesis Testing... yes, it's here, too!

When doing linear regression, it not sufficient to simply "get" a correlation coefficient or the slope of the regression line. These are numbers that are relatively easy to calculate (although you may not agree!). Statisticians are interested in whether these values are statistically significant: i.e., are they significantly different from zero?

As you may imagine, we have ways to determine the significance of *r* or *b,* just as we had ways of determining whether other sample statistics were significant. You guessed it – hypothesis testing!!

Hypothesis Testing and the Correlation Coefficient

Whenever we test hypotheses, we examine the relationship between a sample statistic (such as $\bar{x}$) and its corresponding population parameter (in this case, μ). We do a similar thing in testing the significance of the correlation coefficient, *r*. The corresponding population parameter is ρ, pronounced "rho." It is the population correlation coefficient. The main question is this: do our sample data indicate that ρ is statistically different from zero? This question gives us our hypotheses.

Let's test for the correlation coefficient in our new home square footage vs. cost regression example at the 0.05 level.

Step 1: State the Hypotheses

As the paragraph above indicates, these are our hypotheses:

H_0: $\rho = 0$ (claim)

H_1: $\rho \neq 0$

Step 2: **Find the Critical Values**

This will be a t-test, so we will use our "regular" t-table. In this case, the degrees of freedom are n – 2. So, since we had 9 ordered pairs, we have df = 9 – 2 = 7. Given our alternative hypothesis, we have a two-tailed test, and from our t-table we can see that the critical values of t are ± 2.447. Thus, we will reject H_0 if our test statistic is less than -2.447 or greater than +2.447.

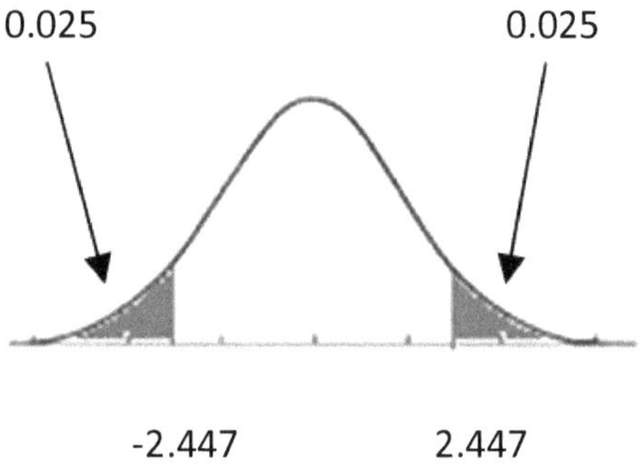

Step 3: **Compute the test statistic**

For this situation, the t-test is different from what you're accustomed to seeing: $t = r\sqrt{\frac{n-2}{1-r^2}}$, where

r is the correlation coefficient, *n – 2* is the degrees of freedom, and r^2 is the coefficient of determination. In this example, $r = 0.9659$, $n - 2 = 7$, and $r^2 = (r)^2 = (0.9659)^2 = 0.93296$. Hence,

$t = r\sqrt{\frac{n-2}{1-r^2}} = (0.9659)\sqrt{\frac{9-2}{1-0.93296}} = 9.8699$. Clearly this value of t is greater than our critical

t = +2.447, well into the upper tail.

Step 4: **Make the Decision**

Again, since 9.8699 > 2.447, we reject the null hypothesis. (By the way, the p-value is $2(1.166 \times 10^{-5}) = 2.232 \times 10^{-5}$, well less than our α of 0.05!!)

Step 5: **State the Conclusion**

Since we reject the null hypothesis, we can state there is a statistically significant correlation between new home square footage and cost.

Hypothesis Testing and the Slope of the Regression Line

Now we will examine the relationship between a calculated slope of the regression line, *b*, and its corresponding population parameter, β, pronounced "beta." Again, the question is this: do our sample data indicate that β is statistically different from zero? This question gives us our hypotheses.

Let's return to our example about new home square footage vs. cost regression example. We will again use the 0.05 level.

Step 1: State the Hypotheses

As the paragraph above indicates, these are our hypotheses:

H_0: $\beta = 0$ (claim)

H_1: $\beta \neq 0$

Step 2: Find the Critical Values

This will be another t-test with df = n – 2. Our critical values of *t* are ± 2.447. Thus, as before, we will reject H_0 if our test statistic is less than -2.447 or greater than +2.447.

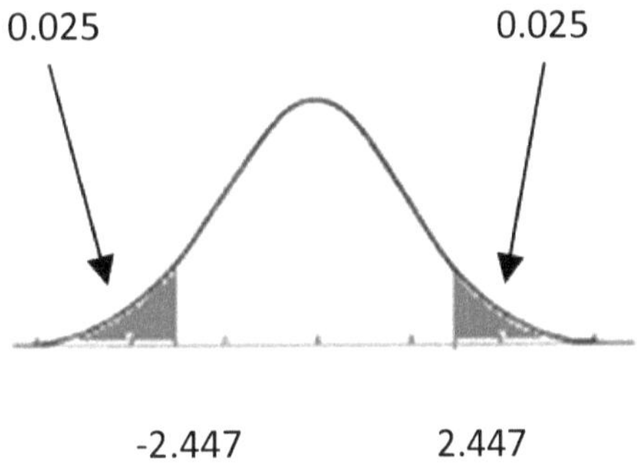

Step 3: Compute the test statistic

This t-test is also different from what you're accustomed to seeing: In this case, $t = \frac{b-\beta}{s_b}$, where s_b represents the standard error of β which can be calculated as follows:

$$s_b = \sqrt{\frac{\frac{\sum(y-\hat{y})^2}{n-2}}{\sum(x-\bar{x})^2}}$$

. While this appears foreboding, we already have most of the required data. The numerator (of the numerator) is the sum of the squared residuals. These can be located in L4 in the TI screenshot on page 5. Simply find the sum of the squared residuals by doing STAT → CALC → 1-Var Stats → 2nd 4 (to get L4). The sum you want is $\sum x^2 = 4194.336714$ (note: the calculator doesn't know what it's summing, so don't be bothered by the x^2 – we have the residuals in the List and x^2 simply means the squared residuals, in this case).

We don't have the denominator, $\sum(x-\bar{x})^2$ – yet! That may be found vary easily. Recall that the x-values are in L1, and that we have already found that $\bar{x} = 32.556$. To get $\sum(x-\bar{x})^2$, we will define L6 as L1 – 32.556. When you hit enter the list will populate. To get the sum of the squares of these entries, simply repeat what we did above: STAT → CALC → 1-Var Stats → 2nd 6 (to get L6). You should see the following in list 6. To the right is a screenshot of the 1-VAR STATS for L6.

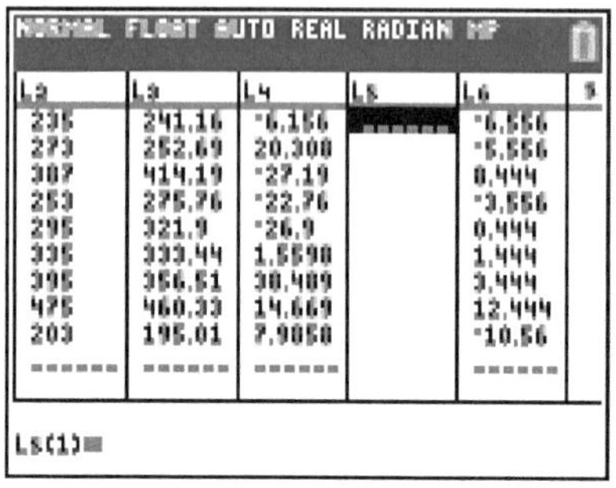

L6 contains $x - \bar{x}$.

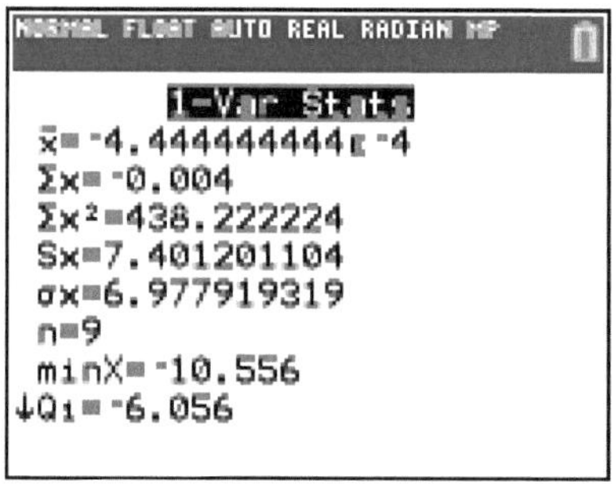

1-VAR STATS for L6. We want $\sum x^2$ $(it's\ really\ \sum(x-\bar{x})^2)$.

Now we're ready to find the standard error of b, s_b.

We know that $\sum(y-\hat{y})^2 = 4194.336714,\ n-2 = 7, and\ \sum(x-\bar{x})^2 = 438.22224.$

$$\text{So, } s_b = \sqrt{\frac{\frac{\sum(y-\hat{y})^2}{n-2}}{\sum(x-\bar{x})^2}} = \sqrt{\frac{\frac{4194.336714}{7}}{438.22224}} = \sqrt{\frac{599.190959}{438.22224}} = \sqrt{1.367322} = 1.169325.$$

Now we can find our test statistic, t: $t = \dfrac{b-\beta}{se(\beta)} = \dfrac{11.539 - 0}{1.169325} = 9.868.$

Step 4: Make the Decision

Clearly this value exceeds our critical t-value, so we reject the null hypothesis. (By the way, the p-value is $2(1.16756 \times 10^{-5}) = 2.335126 \times 10^{-5}$, well less than our α of 0.05!!)

Step 5: State the Conclusion

Since we reject the null hypothesis, we can state that the slope of our regression line is statistically different from zero!! Whew!!

Elaine A. Previte and Dominic Nocera
May 2019

MTH 251	**Names** ______________________
GW – Types of Data	______________________
	______________________

1. Classify each of the following variables considered in the study as qualitative or quantitative.

   a. Nationality

   b. Number of children

   c. Household income in the previous year

   d. Level of education

   e. Daily intake of whole grains (measured in grams per day)

   f. Number of defective items in a lot.

   g. College major of each student in a class.

   h. Salaries of CEOs of oil companies.

   i. Gender of each employee at a company.

   j. Ages of employees at a company.

   k. Method of payment (cash, check, credit card).

2. Classify each as nominal-level, ordinal-level, interval-level, or ratio-level measurement.

   a. Number of pages in your statistics book.

   b. Rankings of PGA golfers.

   c. Weights of air conditioners.

   d. Temperatures inside 10 cars on an August day.

   e. Marital status of patients in a physician's waiting room.

   f. Number of snack and soft drink vending machines in E-Building.

   g. Class rank (Freshman, Sophomore, Junior, Senior).

This page intentionally left blank.

MTH 251 **Names** ____________________

GW – Descr/Infer/Sampling ____________________

____

1. In each of these statements, tell whether descriptive or inferential statistics have been used.

a. Nine out of ten on-the-job fatalities are men (Source: USA TODAY Weekend).

b. The median household income for millennials was $35,592 in 2015 (Source: US Census).

c. Drinking decaffeinated coffee can raise cholesterol levels by 7% (Source: American Heart Association).

d. The national average health care expenditure per capita in the US was $9,403 in 2014 (Source: The World Bank).

e. Experts say that mortgage rates may soon increase.

2. Classify each sample as random, systematic, stratified, or cluster. State why you made your choice.

a. In a large school district, all teachers from two buildings are interviewed to determine whether they believe the students have less homework now than in previous years.

b. Every seventh customer entering a shopping mall is asked to select his or her favorite store.

c. College professors at a large university are selected using random numbers to determine annual salaries.

d. Every 100th hamburger manufactured is checked to determine its fat content.

e. Postal workers in a large city are divided into four groups according to gender and according to whether they walk or ride on their routes. Then 10 are selected from each group and interviewed to determine whether they have been bitten by a dog in the last year.

This page intentionally left blank.

MTH 251	**Names** ________________________
GW – Organizing Data	________________________
	________________________

re ages of US Presidents at their inaugerations, from youngest ot oldest. Use this data to answer the following questions.

sident	Age at Inauguration
odore Roosevelt	42 years, 10 months, 19 days
n Fitzgerald Kennedy	43 years, 7 months, 23 days
liam Jefferson Clinton	46 years, 5 months, 7 days
sses Simpson Grant	46 years, 10 months, 11 days
ack Hussein Obama	47 years, 5 months, 17 days
ver Cleveland (1st term)	47 years, 11 months, 15 days
nklin Pierce	48 years, 3 months, 10 days
es Abram Garfield	49 years, 3 months, 14 days
es Knox Polk	49 years, 4 months, 3 days
lard Fillmore	50 years, 6 months, 3 days
n Tyler	51 years, 0 months, 9 days
vin Coolidge	51 years, 1 month, 0 days
nklin Delano Roosevelt	51 years, 1 month, 5 days
liam Howard Taft	51 years, 5 months, 18 days
ster Alan Arthur	51 years, 11 months, 15 days
aham Lincoln	52 years, 0 months, 21 days
es Earl Carter	52 years, 3 months, 20 days
liam McKinley	54 years, 1 month, 5 days
tin Van Buren	54 years, 3 months, 0 days
herford Birchard Hayes	54 years, 5 months, 1 day
rge Walker Bush	54 years, 6 months, 15 days
bert Clark Hoover	54 years, 6 months, 23 days

Lyndon Baines Johnson	55 years, 2 months, 27 days
Warren Gamaliel Harding	55 years, 4 months, 3 days
Benjamin Harrison	55 years, 6 months, 13 days
Grover Cleveland (2nd term)	55 years, 11 months, 15 days
Richard Milhous Nixon	56 years, 0 months, 12 days
Woodrow Wilson	56 years, 2 months, 5 days
Andrew Johnson	56 years, 3 months, 18 days
George Washington	57 years, 2 months, 9 days
John Quincy Adams	57 years, 7 months, 22 days
Thomas Jefferson	57 years, 10 months, 20 days
James Madison	57 years, 11 months, 17 days
James Monroe	58 years, 10 months, 5 days
Harry S Truman	60 years, 11 months, 5 days
Gerald Rudolph Ford	61 years, 0 months, 22 days
John Adams	61 years, 4 months, 5 days
Andrew Jackson	61 years, 11 months, 18 days
Dwight David Eisenhower	62 years, 3 months, 7 days
Zachary Taylor	64 years, 3 months, 10 days
George Herbert Walker Bush	64 years, 6 months, 28 days
James Buchanan	65 years, 10 months, 10 days
William Henry Harrison	68 years, 0 months, 24 days
Ronald Wilson Reagan	69 years, 11 months, 11 days
Donald John Trump	70 years, 7 months, 7 days

1. Create a frequency distribution containing 6 classes, beginning with the youngest age (***do not*** round years).

Classes	Tally	Frequency	Relative Frequency
Total			

2. Use the data to create a frequency histogram and a relative frequency histogram. How do they compare?

3. Use the raw data to create a stem-and-leaf display.

4. Find the five-number summary and draw the box-and-whiskers display.

   Low = _____ Q_1 = _____ Median = ______ Q_3 = _____ High = _____

   Box-and-whiskers:

MTH 251	**Name** ____________________
GW – Measures of Center	____________________
	____________________

1. Calculate the mean, median and mode of the following grade point averages:

   3.2 2.5 2.8 3.7 2.8 2.0

2. Calculate the mean for samples where

   a. $n = 10$, $\sum x = 85$ b. $n = 16$, $\sum x = 400$

3. Let's assume you have the following sample: 0, 6, 7, 7, 8, 8, 8, 9, 9, 10.

   a. Find the mean and median of the original sample.

   b. Now replace the 0 with 4. How do the mean and median change?

   c. Now replace the 10 with 100. What happens now?

   d. Explain why there were changes to the mean and/or median as you made the changes required in parts b and c.

This page intentionally left blank.

.

MTH 251 **Name** ________________________

GW – Mean & Standard Deviation ________________________

____

1. Observe the three data sets illustrated below. For each set, compute the sample mean. Question: If you were to use one descriptive measure to discuss these data sets, do you think the sample mean would be the best such measure? Why or why not?

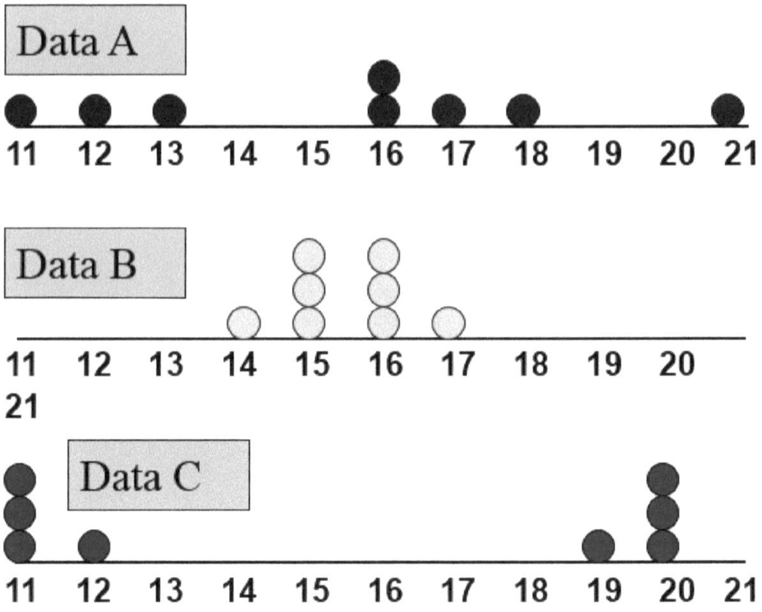

2. Compute the sample standard deviation for each data set. Question: Does the addition of this descriptive measure enhance your discussion about the data sets?

Data Set A

x	$x - \bar{x}$	$(x - \bar{x})^2$

Data Set B

x	$x - \bar{x}$	$(x - \bar{x})^2$

Data Set C

x	$x - \bar{x}$	$(x - \bar{x})^2$

MTH 251	**Name** _________________________
GW – Measures of Dispersion	_________________________
	_________________________

1. Calculate the range, variance, and standard deviation for the following samples:

a. 4, 2, 1, 0, 1

x	$x - \bar{x}$	$(x - \bar{x})^2$

b. 5, 5,5, 5, 5, 5, 5

x	$x - \bar{x}$	$(x - \bar{x})^2$

2. Use the shortcut formula to find the sample standard deviation given the following information:
Note: $s^2 = \frac{n\sum x^2 - (\sum x)^2}{n(n-1)}$ $and\ s = \sqrt{\frac{n\sum x^2-(\sum x)^2}{n(n-1)}}$.

a. n = 10, $\sum x = 504$, $\sum x^2 = 39612$

b. n = 5, $\sum x = 55$, $\sum x^2 = 749$

This page intentionally left blank.

MTH 251 **Names** ____________________________

Chebyshev's Theorem, Empirical Rule & z-scores ____________________________

____

1. The average earnings of year-round full-time workers 25–34 years old with a bachelor's degree or higher were $58,500 in 2003. If the standard deviation is $11,200, and the data appear to be bell-shaped, what can you say about the percentage of these workers who earn

   a. Between $47,300 and $69,700?

   b. More than $80,900?

   c. How likely is it that someone earns more than $100,000?

2. Americans spend an average of 3 hours per day online. If the standard deviation is 32 minutes, find the range in which at least 88.89% of the data will lie. Use Chebyshev's theorem.

3. If the average number of vacation days for a selection of various countries has a mean of 29.4 days and a standard deviation of 8.6, find the *z* scores for the average number of vacation days in each of these countries.

Canada	26 days
Italy	42 days
United States	13 days

6. The average teacher's salary in a particular state is $54,166. If the standard deviation is $10,200, find the salaries corresponding to the following *z* scores.

a. 2	c. 0	*e.* 2.5
b. −1	*d.* −1.6	*f.* -3.0

MTH 251	Names ____________________
GW – Covariance & Correlation	____________________
	____________________

1. The following is a set of data from a sample of n items:

x	8	5	8	3	7	10	12	4	9	15	18
y	21	16	25	10	20	32	34	14	29	46	50

   a. Draw a scatterplot of the data (remember, these are ordered pairs). What type of relationship is there?

   b. Compute the covariance: $COV(\mathrm{x}, \mathrm{y}) = \frac{\sum(x-\bar{x})(y-\bar{y})}{n-1}$

   c. Compute the coefficient of correlation: $r = \frac{COV(x,y)}{s_x \cdot s_y}$

   d. How strong is the relationship between x and y? Explain.

2. A random sample of 11 statistics students produced the following data, where x is the third exam score out of 80, and y is the final exam score out of 200.

   a. Plot the data on the coordinate system below (which is the independent variable and which is the dependent variable?).

   b. What type of correlation does there appear to be? Is it strong or weak?

   c. Compute the correlation coefficient. Is the relationship strong or weak as you predicted?

x (third exam score)	y (final exam score)
65	175
67	133
71	185
71	163
66	126
75	198
67	153
70	163
71	159
69	151
69	159

MTH 251 **Names** ____________________

Basic Probability ____________________

____

1. Suppose a card is drawn from a standard deck of cards. Find the probability that the following cards are chosen:

    a. An ace

    b. A diamond

    c. An ace of diamonds

    d. A face card or a three

    e. A 4 or a 6

    f. A 6 or a club

    g. A heart or a club

    h. A red king

    i. A red card or a 7

    j. A black card and a 10

    k. A king given it's a face card

    l. A 10 given it's a black card

2. a. Are the events "4" or "6" mutually exclusive? Why or why not? Independent? Why or why not?

    b. Are the events "king" and "black card" mutually exclusive? Why or why not? Independent? Why or why not?

    c. Are the events "10" and "black card" mutually exclusive? Why or why not? Independent? Why or why not?

This page intentionally left blank.

MTH 251 **Name** ____________________

Conditional Probability ____________________

____

1. A recent survey asked 100 Bristol Business Administration majors if they were planning to transfer to a Massachusetts state university or a private university. The results of the survey are shown.

Transfer:	State	Private	Total
Male	15	18	33
Female	25	42	67
Total	40	60	100

Find these probabilities.

a. The respondent is transferring to a state university, given that the respondent was a female.

b. The respondent was a male, given that the respondent is transferring to a private university.

c. Are the events "state university" and "female" independent? Why or why not?

2. Three cards are drawn from a deck *without* replacement. Find these probabilities.
a. All are jacks.
b. All are clubs.
c. All are red cards.

3. A vase contains 5 red marbles and 3 blue marbles. Two marbles are chosen at random and are not replaced. Find these probabilities. Repeat the problem, this time replacing the marbles.
a. Selecting a red marble and a blue marble.
b. Selecting two blue marbles.
c. Selecting at least one red marble.

This page intentionally left blank.

MTH 251	**Name** ____________________
Counting Rules	____________________
	____________________

1. How many 5-digit zip codes are possible if digits can be repeated?

   If there cannot be repetitions?

2. How many ways can a baseball manager arrange a batting order of 9 players?

3. If a person can select 3 presents from 10 presents under a Christmas tree, how many different combinations are there?

4. How many ways can a dinner patron select 3 appetizers and 2 vegetables if there are 6 appetizers and 5 vegetables on the menu?

5. The composition of the 108th Congress was 51 Republicans, 48 Democrats, and 1 Independent. A committee on aid to higher education is to be formed with 3 Senators to be chosen at random to head the committee. Find the probability that the group of 3 consists of

   a. All Republicans

   b. All Democrats

   c. One Democrat, one Republican, and one Independent

This page intentionally left blank.

MTH 251	**Name** ____________________
Discrete Probability Distributions	____________________
	____________________

Find the mean, variance and standard deviation of each discrete probability distribution.

1. The probabilities that a bakery has a demand for 2, 3, 5, or 7 birthday cakes on any given day are 0.35, 0.41, 0.15, and 0.09, respectively.

2. The probabilities that a customer will rent 0, 1, 2, 3, or 4 DVDs on a single trip to RedBox are 0.15, 0.25, 0.30, 0.25, and 0.05, respectively.

This page intentionally left blank.

MTH 251	**Names** ____________
GW: Binomial Distribution	____________
	____________

1. Given n = 6 trials, p = 0.70, find the following:

   a. P(x = 4)

   b. P(x = 6)

   c. P(x = 0)

   d. P(x < 2)

   e. $P(x \geq 2)$

2. You are to take a 5-question T-F quiz on Econometrics. Find the following probabilities:

   a. P(you get 100 on the quiz)

b. P(you fail* the quiz)

* failing means earning less than 60% on the quiz.

3. Suppose 70% of voters in the BCC community tend to vote Democrat. If you ask the next 4 people you see in the hallway, what is the probability that:

   a. None of them voted Democrat during the last election?

   b. At least one of them voted Democrat?

   c. All of them voted Democrat?

   d. Between 2 and 4 (inclusive) of them voted Democrat?

4. A survey found that one out of five Americans say he or she has visited a doctor in any given month. If 10 people are selected at random, find the probability that exactly 3 will have visited a doctor last month.

5. A recent survey found that 30% of teenage consumers receive their spending money from part-time jobs. If 5 teenagers are selected at random, find the probability that at least 3 of them will have part-time jobs.

MTH 251	**Names** ______________________
GW – The Normal Curve	______________________
	______________________

1. Find the following probabilities. Hint: draw the pictures!

a) P(z > 1.645)

b) P(z < 0.67)

c) P(|z| > 1.96)

2. Find and illustrate the z-score such that:

a) 12.3% of the area under the curve lies to the left of it	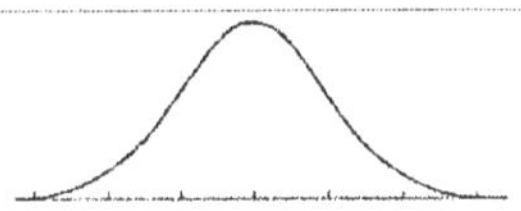
b) 82.12% of the area under the curve lies to the right of it	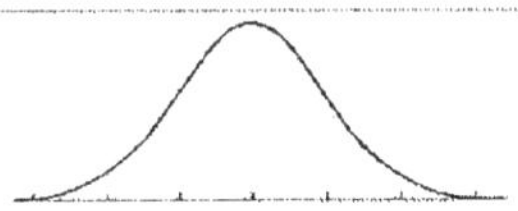
c) 48% of the area under the curve is symmetric about the mean	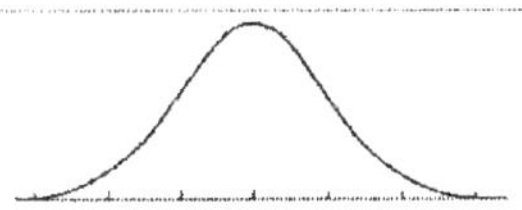

3. Suppose a milk producer has a daily production, x, that is approximately normally distributed with a mean of 200,000 gallons and a standard deviation of 20,000 gallons. Corporate management will give a bonus if daily production exceeds the 90th percentile. How many gallons need to be produced so that the bonus is awarded?

4. A certain investment account pays 1.5%, on average, with a standard deviation of 0.2%. Assuming the interest rates are normally distributed, find the probabilities that:

   a) The account pays between 1.3% and 1.6%.

   b) The account pays more that 1.4%.

   c) The account pays less than 1.5%

MTH 251 **Names** ______________________

GW –Sampling Distributions of Sample Mean ______________________

____

1. A manufacturer of car batteries claims that the distribution of the life of its best battery has a mean of 54 months with a standard deviation of 6 months. In recent months, the manufacturer received many complaints that its batteries have died much earlier than advertised. Suppose you want to test the claim by sampling 50 batteries. Test the claim that the batteries died in 52 months or less.

2. According to a recent study, full-time college students majoring in engineering spend an average of 19 hours per week studying (http://www.schools.com/articles/how-much-do-college-students-study.html). Let's assume a standard deviation of 2 hours, and that study times are normally distributed.

a) If an engineering major is selected at random, find the probability that s/he spends at least 20 hours studying each week. Show all work.

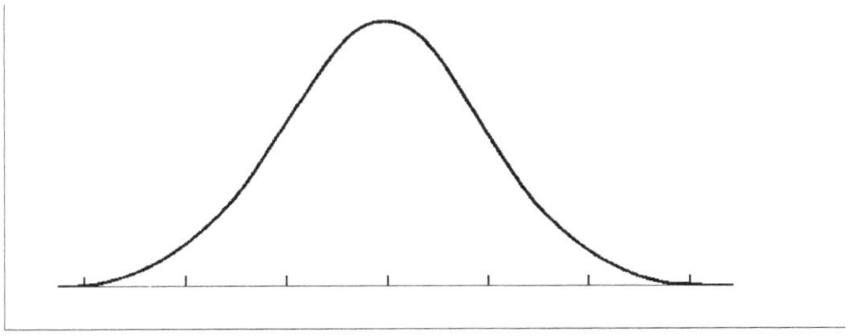

b) If a sample of 36 engineering majors is selected at random, find the probability that the group's mean amount of time studying is at least 20 hours studying each week. Show all work.

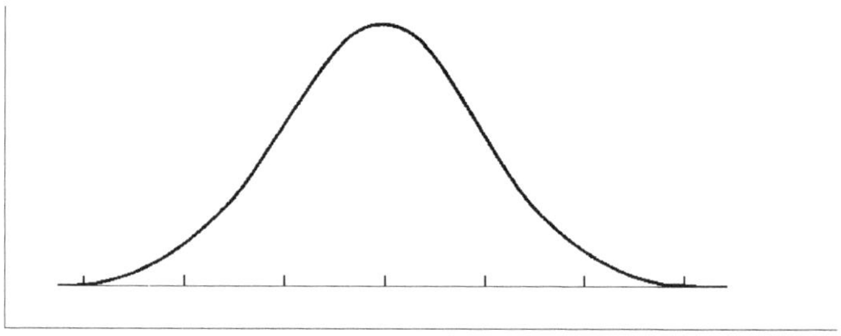

3. The average age of doctors in a certain hospital is 43.0 years old. Suppose the distribution of ages is normal and has a standard deviation of 8.0 years (assume that the variable is normally distributed). If 25 doctors are chosen at random for a committee, find the probability that the average age of those doctors is:

a) less than 43.8 years

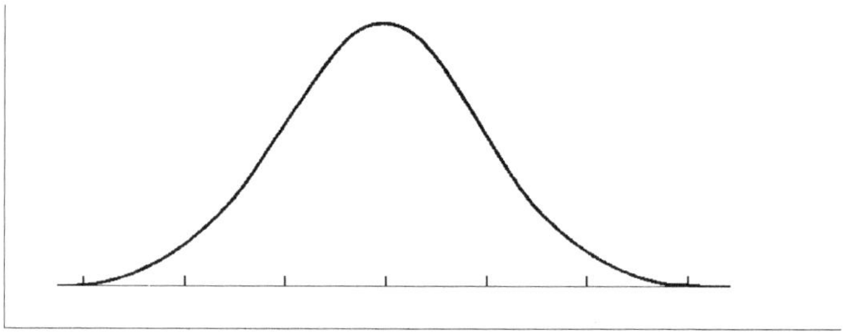

b) more than 43.8 years

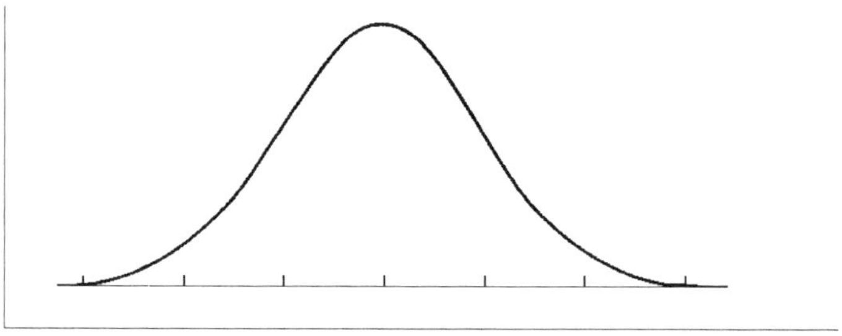

c) between 43 and 43.8 years

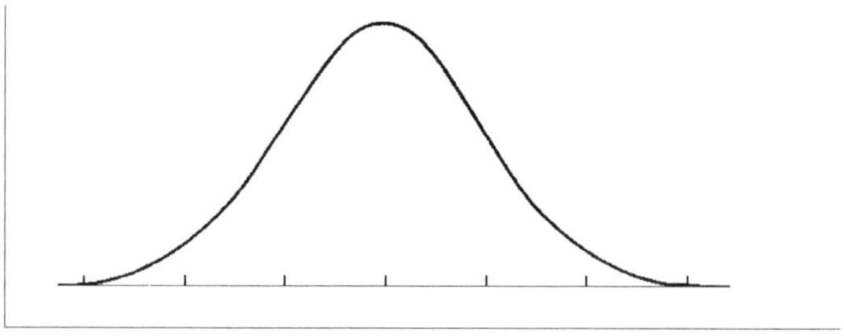

MTH 251	**Names** ________________________
GW –Sampling Distributions of Sample Proportion	________________________
	________________________

1. According to a recent survey, 15% of U.S. adults admitted they are "cord cutters," i.e., they canceled the cable/satellite TV service they once subscribed to. In a random sample of 500 U.S. adults, let $\hat{p}$ represent the proportion who are "cord cutters."

   a. Compute the probability that $\hat{p}$ is less than 0.12.

   b. Compute the probability that $\hat{p}$ is greater than 0.10.

2. According to a recent poll of U.S. adults about 60% work during their summer vacation. Assume that the true proportion of all US. adults who work during summer vacation is p = .6. Now consider a random sample of 500 U.S. adults.

   a. What is the probability that between 55% and 65% of the sampled adults work during summer vacation?

   b. What is the probability that over 75% of the sampled adults work during summer vacation?

3. According to an OHSA 2014 report, a hospital is one of the most dangerous places to work. A major cause of injuries that result in missed work was overexertion. Almost half (48%) of the injuries that result in missed work were due to overexertion. Consider a random sample of 100 hospital workers. What is the probability that the sample proportion is less than 0.40.

This page intentionally left blank.

MTH 252 **Names** ____________________________

Confidence Intervals (μ, σ known) ____________________________

1. A survey of 64 individuals who passed the seven exams and obtained the rank of Fellow in the actuarial field finds the average salary to be $150,000. If the standard deviation for the population is $15,000, construct a 95% confidence interval for all Fellows.

Step 1: What is the population of interest?

Step 2: What is the sample size? n = ________

Step 3: What is the confidence level? ________

What is α? α = _____

Step 4: What is the sample mean (point estimate)? $\bar{x} =$ _____

Step 5: What is the standard error of the sampling distribution? $\sigma_{\bar{x}} = \frac{\sigma}{\sqrt{n}} = \frac{\quad}{\sqrt{\quad}} =$ _____

Step 6: Which value of z will you use? z = _________

Step 7: Compute the confidence interval estimate. $\bar{x} \pm z \cdot \frac{\sigma}{\sqrt{n}}$

Step 8: Interpret the result.

2. A study of 35 golfers showed that their average score on a particular course was 92. The standard deviation of the population is 5. Find the 90% confidence interval of the true mean for all golfers at that golf course.

Step 1: What is the population of interest?

Step 2: What is the sample size? n = ________

Step 3: What is the confidence level? ________

What is α? α = _____

Step 4: What is the sample mean (point estimate)? $\bar{x} =$ ____

Step 5: What is the standard error of the sampling distribution? $\sigma_{\bar{x}} = \frac{\sigma}{\sqrt{n}} = \frac{\quad}{\sqrt{\quad}} =$ ____

Step 6: Which value of z will you use? z = _________

Step 7: Compute the confidence interval estimate. $\bar{x} \pm z \cdot \frac{\sigma}{\sqrt{n}}$

Step 8: Interpret the result.

MTH 252 **Names** ______________________

Confidence Intervals (for μ; σ unknown) ______________________

1. The approximate costs (in thousands of dollars) for a 30-second spot for various cable networks in a random selection of cities are shown below. Estimate the true population mean cost for a 30-second advertisement on cable network with 90% confidence.

14	55	165	13	9	54	15	73
66	55	23	41	30	78	150	

Step 1: What is the population of interest?

Step 2: What is the sample size? n = ________

How many degrees of freedom are there? df = __________

Step 3: What is the confidence level? ________

What is α? α = _____

Step 4: What is the sample mean (point estimate)? $\bar{x} =$ _____

Step 5: What is the standard error of the sampling distribution? $\sigma_{\bar{x}} = \frac{s}{\sqrt{n}} = \frac{\quad}{\sqrt{\quad}} =$ _____

Step 6: Which value of t will you use? t = _________

Step 7: Compute the confidence interval estimate. $\bar{x} \pm t \cdot \frac{s}{\sqrt{n}}$

Step 8: Interpret the result.

2. For a group of 22 college football players, the mean heart rate after a morning workout session was 86 beats per minute, and the standard deviation was 5. Find the 90% confidence interval of the true mean for all college football players after a workout session. If a coach did not want to work his team beyond its capacity, what maximum value should he use for the mean number of heartbeats per minute?

Step 1: What is the population of interest?

Step 2: What is the sample size? n = ________

How many degrees of freedom are there? df = __________

Step 3: What is the confidence level? ________

What is α? α = _____

Step 4: What is the sample mean (point estimate)? $\bar{x}$ = ____

Step 5: What is the standard error of the sampling distribution? $\sigma_{\bar{x}} = \frac{s}{\sqrt{n}} = \frac{\quad}{\sqrt{\quad}} =$ ____

Step 6: Which value of t will you use? t = _________

Step 7: Compute the confidence interval estimate. $\bar{x} \pm t \cdot \frac{s}{\sqrt{n}}$

Step 8: Interpret the result.

MTH 252 **Names** ______________________

GW – CI for Population Proportions

____

1. A U.S. Travel Data Center survey conducted for *Better Homes and Gardens* of 1500 adults found that 39% said that they would take more vacations this year than last year. Find the 95% confidence interval for the true proportion of adults who said that they will travel more this year.

2. A random sample of 205 college students was asked if they believed that places could be haunted, and 65 of them responded yes. Estimate the true proportion of college students who believe in the possibility of haunted places with 99% confidence. According to *Time* magazine, 37% of Americans believe that places can be haunted.

3. A sample of 400 Americans aged 25 to 29 yielded 190 who were unmarried. Find the90% confidence interval of the true proportion of Americans in this age group who are unmarried.

This page intentionally left blank.

MTH 252 **Names** ____________________

GW – CI - Sample Sizes ____________________

____

1. A university dean of students wishes to estimate the average number of hours students spend doing homework per week. The standard deviation from a previous study is 6.2 hours. How large a sample must be selected if he wants to be 99% confident of finding whether the true mean differs from the sample mean by 1.5 hours?

2. It is desired to estimate the mean number of chocolate chips per cookie for a large national brand. How many cookies would have to be sampled to estimate the true mean number of chips per cookie within 2 chips with 98% confidence? Assume $\sigma = 10.1$ chips.

3. A medical researcher wishes to determine the percentage of females who take vitamins.He wishes to be 99% confident that the estimate is within 2 percentage points of the true proportion. A recent study of 180 females showed that 25% took vitamins.
   a. How large should the sample size be?
   b. If no estimate of the sample proportion is available, how large should the sample be?

This page intentionally left blank.

Math 252 **Names** ______________________

GW – Intro to Hypothesis Testing ______________________

____

The YouTube video discussed whether we should believe Sam's claim that his bowling average is (at least) 150, despite the fact that he averaged 40 in his three games with you. Let's assume that the standard deviation of all Sam's games is 75. At the 0.05 level of significance, do you accept Sam's claim? What if his standard deviation is 125?

H_0:

H_1: (which one is the claim?)

μ =	Since σ is known, the test statistic is:
n =	$z = \frac{\bar{x} - \mu}{\sigma / \sqrt{n}}$
$\bar{x}$ =	
σ =	
α =	

Draw rejection region & label critical value(s):

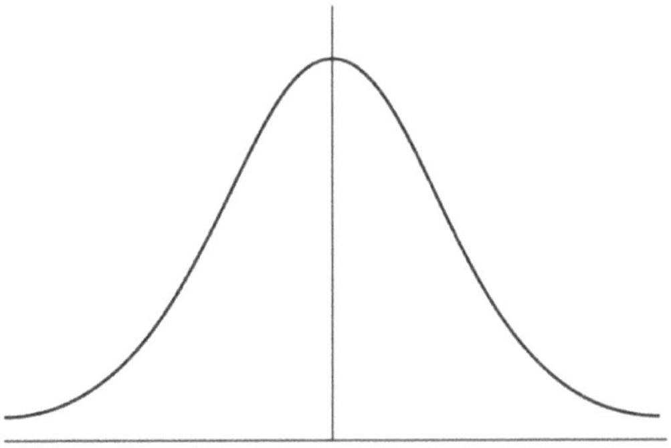

Calculations:

Decision:

Conclusion:

This page intentionally left blank.

Math 252	Names ____________________
	____________________
GW - Hypothesis Testing for μ, σ known	____________________

1. A researcher reports that the average salary of assistant professors is more than \$42,000. A sample of 30 assistant professors has a mean salary of \$43,260. At $\alpha = 0.05$, test the claim that assistant professors earn more than \$42,000 per year. The standard deviation of the population is \$5230.

H_0:

H_1:

(which one is the claim?)

$\mu =$ $n =$ $\bar{x} =$ $\sigma =$ $\alpha =$	Since σ is known, the test statistic is: $z = \dfrac{\bar{x} - \mu}{\sigma / \sqrt{n}}$

Draw rejection region & label critical value(s):

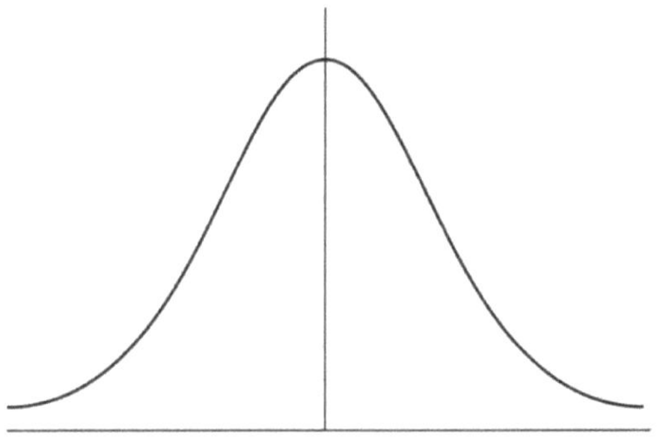

Calculations:

Decision:

Conclusion:

2. A cold capsule is supposed to contain 10 grains of aspirin, with a standard deviation of 1.4 grains. A random sample of 100 tablets yielded a mean of 10.2 grains of aspirin. Can it be concluded that μ is different from 10? Use $\alpha = 0.05$.

H_0:

H_1: (which one is the claim?)

μ =	Since σ is known, the test statistic is:
n =	$z = \frac{\bar{x} - \mu}{\sigma / \sqrt{n}}$
$\bar{x}$ =	
σ =	
α =	

Draw rejection region & label critical value(s):

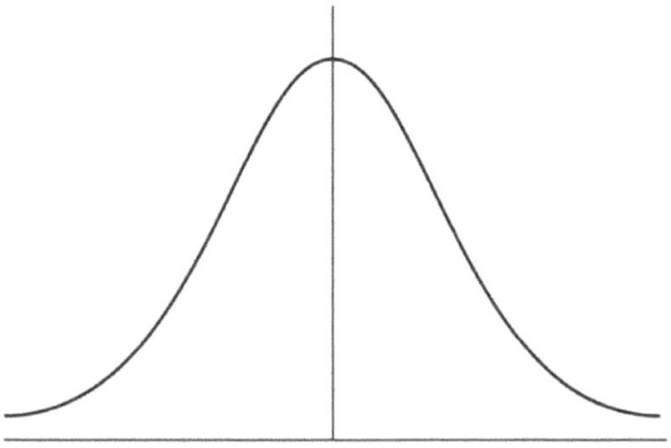

Calculations:

Decision:

Conclusion:

3. A psychologist is studying the effect of chronic alcohol abuse on memory. In this experiment, a memory test is used; scores on this test are normally distributed with a mean of 50 and a standard deviation of 6.
   A sample of 22 alcohol abusers has an average score of 47. Is there evidence to conclude there is memory impairment among alcohol abusers? Use $\alpha = 0.01$.

H_0:

H_1:

(which one is the claim?)

μ =

n =

$\bar{x}$ =

σ =

α =

Since σ is known, the test statistic is:

$$z = \frac{\bar{x} - \mu}{\sigma / \sqrt{n}}$$

Draw rejection region & label critical value(s):

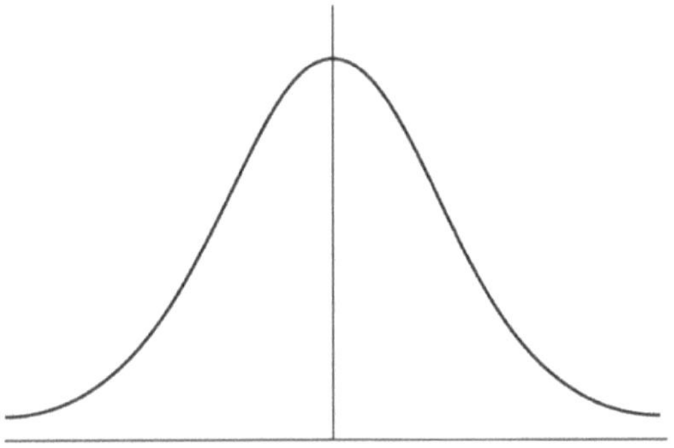

Calculations:

Decision:

Conclusion:

4. A researcher would like to assess the "miraculous" claims of improvement made in a TV advertisement about a phonetic reading instruction package. It is known that the scores on a standardized reading test for nine-year-olds is normally distributed with a mean of 45 and a standard deviation of 10.
A random sample of 25 eight-year-olds is given the reading package for a year. At age nine, these children are given the standardized reading test. Their mean score is 48.6. Does this indicate that the reading package is effective? Use $\alpha = 0.05$.

H_0: H_1:	(which one is the claim?)

μ = n = $\bar{x}$ = σ = α =	Since σ is known, the test statistic is: $z = \dfrac{\bar{x} - \mu}{\sigma / \sqrt{n}}$

Draw rejection region & label critical value(s):	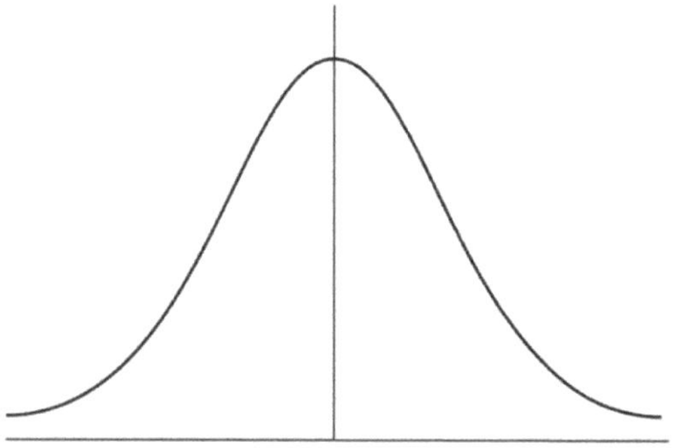

Calculations:

Decision:

Conclusion:

MTH 252	**Names** ____________________
	____________________
GW - Hypothesis Testing for μ, σ unknown	____________________

1. A medical investigation claims that the average number of infections per week at a hospital in southwestern Pennsylvania is 16.3. A random sample of 10 weeks had a mean number of 17.7 infections. The sample standard deviation is 1.8. Is there enough evidence to reject the investigator's claim at $\alpha = 0.05$?

H_0:

H_1:

(which one is the claim?)

μ =	Since σ is unknown, the test statistic is:
n =	$t = \dfrac{\bar{x} - \mu}{s/\sqrt{n}}$
df =	
$\bar{x}$ =	
s =	
α =	

Draw rejection region & label critical value(s):

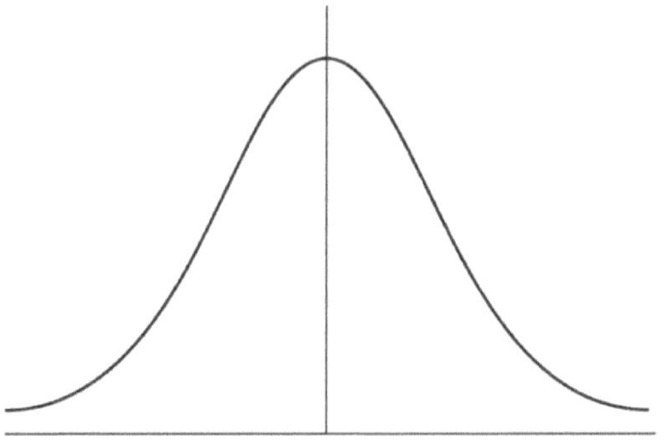

Calculations:

Decision:

Conclusion:

2. An educator claims that the average salary of substitute teachers in school districts in Allegheny County, Pennsylvania, is less than \$60 per day. A random sample of eight school districts is selected, and the daily salaries (in dollars) are shown. Is there enough evidence to support the educator's claim at $\alpha = 0.10$?

   60 56 60 55 70 55 60 55

H_0:

H_1:

(which one is the claim?)

$\mu =$ $n =$ df = $\bar{x} =$ $s =$ $\alpha =$	Since σ is unknown, the test statistic is: $t = \dfrac{\bar{x} - \mu}{s/\sqrt{n}}$

Draw rejection region & label critical value(s):

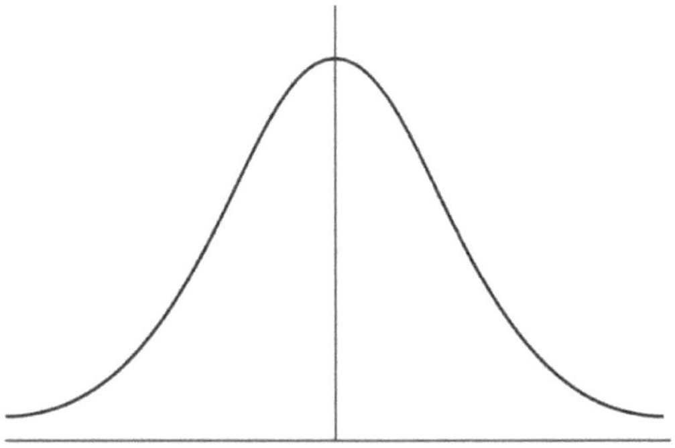

Calculations:

Decision:

Conclusion:

MTH 252	Names ______________________
	______________________
GW - Hypothesis Testing for p	______________________

1. A dietitian claims that 60% of people are trying to avoid transfats in their diets. She randomly selected 200 people and found that 128 people stated that they were trying to avoid transfats in their diets. At $\alpha = 0.05$, is there enough evidence to reject the dietitian's claim?

H_0:

H_1: (which one is the claim?)

p = q = 1 - p x = n = $\hat{p}$ = α =	Since this is about a proportion, the test statistic is: $z = \dfrac{\hat{p} - p}{\sqrt{\dfrac{pq}{n}}}$

Draw rejection region & label critical value(s):	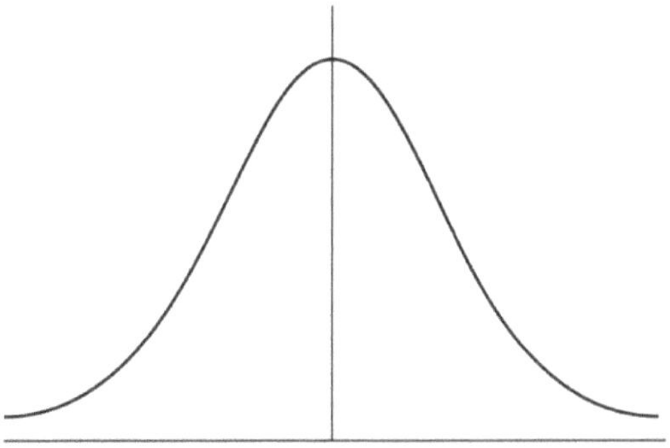

Calculations:

Decision:

Conclusion:

2. A telephone company representative estimates that at most 40% of its customers have call-waiting service. To test this hypothesis, she selected a sample of 100 customers and found that 43% had call waiting. At $\alpha = 0.01$, is there enough evidence to reject the claim?

H_0: H_1:	(which one is the claim?)

p = q = 1 - p x = n = $\hat{p}$ = α =	Since this is about a proportion, the test statistic is: $$z = \frac{\hat{p} - p}{\sqrt{\frac{pq}{n}}}$$

Draw rejection region & label critical value(s):	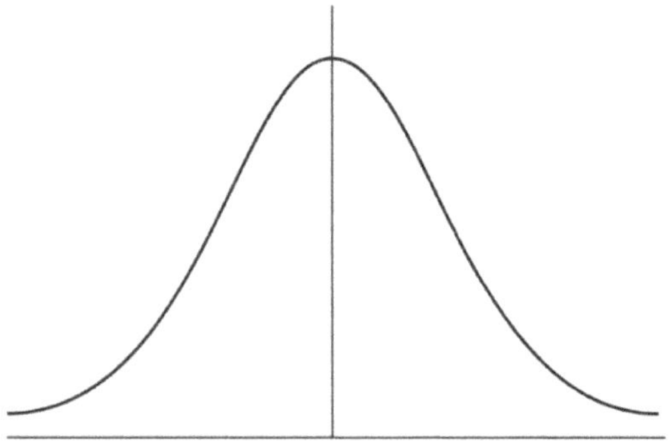

Calculations:

Decision:

Conclusion:

MTH 252	Names ____________________
	____________________
GW - Hypothesis Testing – Two Samples – z-test	____________________

1. A survey found that the average hotel room rate in Boston is $188.42 and the average room rate in Providence is $180.61. Assume that the data were obtained from two samples of 50 hotels each and that the standard deviations of the populations are $15.62 and $14.83, respectively. At α= 0.05, can it be concluded that there is a significant difference in the rates?

H_0:

H_1: (which one is the claim?)

$\mu_1 - \mu_2 =$		Since this is about the difference between two means with known population standard deviations, the test statistic is:
$\sigma_1 =$	$\sigma_2 =$	$$z = \frac{(\bar{x}_1 - \bar{x}_2) - (\mu_1 - \mu_2)}{\sqrt{\frac{\sigma_1^2}{n_1} + \frac{\sigma_2^2}{n_2}}}$$
$\bar{x}_1 =$	$\bar{x}_2 =$	
$n_1 =$	$n_2 =$	
$\alpha =$		

Draw rejection region & label critical value(s):

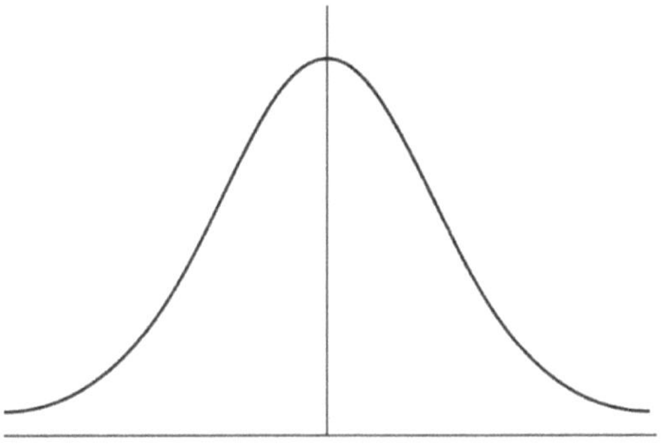

Calculations:

Decision:

Conclusion:

2. It is believed that the average length of "short hospital stays" for men is slightly longer than that for women. A random sample of recent hospital stays for both men and women revealed the following. At $\alpha = 0.01$, is there sufficient evidence to conclude that the average hospital stay for men is longer than the average hospital stay for women?

	Men	**Women**
Sample size	32	30
Sample mean	5.5 days	4.2 days
Population standard deviation	1.2 days	1.5 days

H_0:

H_1:

(which one is the claim?)

$\mu_1 - \mu_2 =$

$\sigma_1 =$ $\sigma_2 =$

$\bar{x}_1 =$ $\bar{x}_2 =$

$n_1 =$ $n_2 =$

$\alpha =$

Since this is about the difference between two means with known population standard deviations, the test statistic is:

$$z = \frac{(\bar{x}_1 - \bar{x}_2) - (\mu_1 - \mu_2)}{\sqrt{\frac{\sigma_1^2}{n_1} + \frac{\sigma_2^2}{n_2}}}$$

Draw rejection region & label critical value(s):

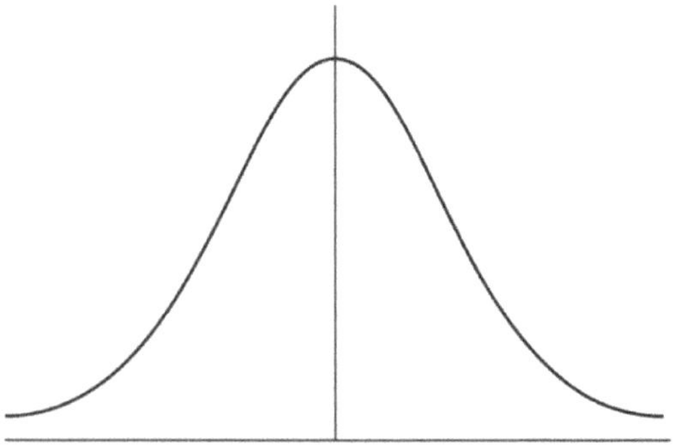

Calculations:

Decision:

Conclusion:

MTH 252 **Names** ____________________

____

GW - Hypothesis Testing - 2 samples – separate variance t-test ____________________

1. The average size of a farm in State A is 235 acres. The average size of a farm in State B is 199 acres. Assume the data were obtained from two samples with standard deviations of 38 and 12 acres, respectively, and sample sizes of 8 and 10, respectively. Can it be concluded at $\alpha = 0.05$ that the average size of the farms in the two counties is different? Assume the populations are normally distributed and that population variances are unequal.

H_0:

H_1:

(which one is the claim?)

$\mu_1 - \mu_2 =$		Since this is about the difference between two means with unequal population variances, the test statistic is:
$s_1 =$	$s_2 =$	$$t = \frac{(\bar{x}_1 - \bar{x}_2) - (\mu_1 - \mu_2)}{\sqrt{\frac{s_1^2}{n_1} + \frac{s_2^2}{n_2}}}$$
$\bar{x}_1 =$	$\bar{x}_2 =$	
$n_1 =$	$n_2 =$	
$\alpha =$	$df = \gamma =$	where $df = \gamma = \dfrac{\left(\frac{s_1^2}{n_1} + \frac{s_2^2}{n_2}\right)^2}{\frac{\left(\frac{s_1^2}{n_1}\right)^2}{n_1 - 1} + \frac{\left(\frac{s_2^2}{n_2}\right)^2}{n_2 - 1}}$

Draw rejection region & label critical value(s):

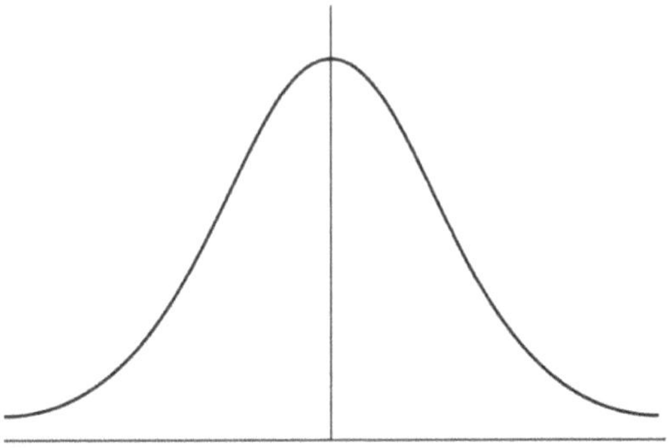

Calculations:

Decision:

Conclusion:

2. Whiting, Indiana, leads the "Top 100 Cities with the Oldest Houses" list with the average age of houses being 66.4 years. Farther down the list resides Franklin, Pennsylvania, with an average house age of 59.4 years. Researchers selected a random sample of 20 houses in each city and obtained the following statistics. At α= 0.05, can it be concluded that the houses in Whiting are older? Assume population variances are unequal.

	Whiting	**Franklin**
Sample mean	62.1 years	55.6 years
Sample Standard deviation	5.4 years	3.9 years

H_0:

H_1:

(which one is the claim?)

$\mu_1 - \mu_2 =$

$s_1 =$ $s_2 =$

$\bar{x}_1 =$ $\bar{x}_2 =$

$n_1 =$ $n_2 =$

$\alpha =$ $df = \gamma =$

Since this is about the difference between two means with unequal population variances, the test statistic is:

$$t = \frac{(\bar{x}_1 - \bar{x}_2) - (\mu_1 - \mu_2)}{\sqrt{\frac{s_1^2}{n_1} + \frac{s_2^2}{n_2}}}$$

where $df = \gamma = \dfrac{\left(\frac{s_1^2}{n_1} + \frac{s_2^2}{n_2}\right)^2}{\frac{\left(\frac{s_1^2}{n_1}\right)^2}{n_1 - 1} + \frac{\left(\frac{s_2^2}{n_2}\right)^2}{n_2 - 1}}$

Draw rejection region & label critical value(s):

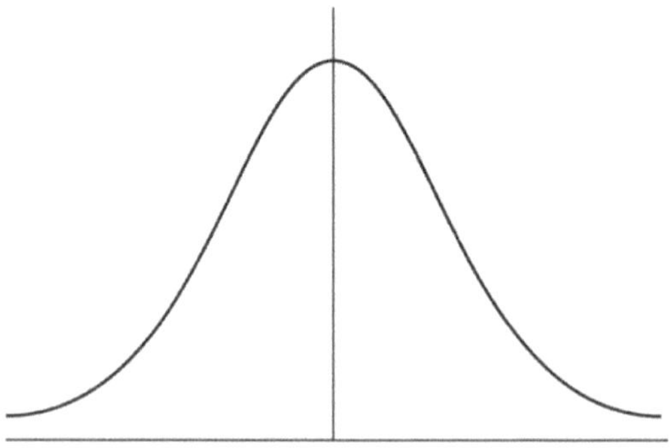

Calculations:

Decision:

Conclusion:

MTH 252 **Names** ____________________

____

GW - Hypothesis Testing - 2 samples – pooled t-test ____________________

1. You and some friends have decided to test the validity of an advertisement by a local pizza restaurant which says it delivers to the dormitories faster than a local branch of a national chain. Both restaurants are located across the street from the college. You define the variable of interest as the delivery time, in minutes, from when the pizza is ordered to when it is delivered. You collect data by ordering 10 pizzas from the local pizza restaurant and 10 pizzas from the national chain at different times. You store the summary statistics in the table below. At the 0.05 level of significance, is there evidence that the mean delivery time for the local pizza restaurant is less than the mean delivery time for the national pizza chain? Assume population variances are equal.

Local	**Chain**
Mean time: 16.7 min	Mean time: 18.88.min
Std. Dev.: 3.1 min	Std. Dev.: 2.87 min

H_0:

H_1:

(which one is the claim?)

$\mu_1 - \mu_2 =$

$s_1 =$ $s_2 =$

$\bar{x}_1 =$ $\bar{x}_2 =$

$n_1 =$ $n_2 =$

$\alpha =$ df $= n_1 + n_2 - 2 =$

Since this is about the difference between two means with equal population variances, the test statistic is:

$$t = \frac{(\bar{x}_1 - \bar{x}_2) - (\mu_1 - \mu_2)}{\sqrt{s_p^2\left(\frac{1}{n_1} + \frac{1}{n_2}\right)}}$$

where $s_p^2 = \frac{(n_1-1)s_1^2 + (n_2-1)s_2^2}{n_1+n_2-2}$

Draw rejection region & label critical value(s):

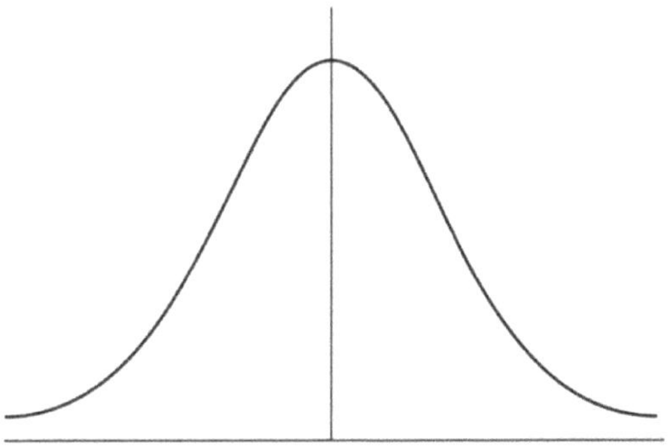

Calculations: Decision: Conclusion:

2. A bank with a branch located in a commercial district of a city has the business objective of developing an improved process for serving customers during the noon-to-1 P.M. lunch period. Management decides to first study the waiting time in the current process. The waiting time is defined as the number of minutes that elapses from when the customer enters the line until he or she reaches the teller window. Data are collected from a random sample of 15 customers and are:

4.21	5.55	3.02	5.13	4.77	2.34	3.54	3.20
4.50	6.10	0.38	5.12	6.46	6.19	3.79	

Suppose that another branch, located in a residential area, is also concerned with improving the process of serving customers in the noon-to-1 P.M. lunch period. Data are collected from a random sample of 15 customers and stored These data are:

9.66	5.90	8.02	5.79	8.73	3.82	8.01	8.35
10.49	6.68	5.64	4.08	6.17	9.91	5.47	

Assuming that the population variances from both banks are equal, is there evidence of a difference in the mean waiting time between the two branches? (Use $\alpha = 0.05$.)

H_0:

H_1:

(which one is the claim?)

$\mu_1 - \mu_2 =$

$s_1 =$ $s_2 =$

$\bar{x}_1 =$ $\bar{x}_2 =$

$n_1 =$ $n_2 =$

$\alpha =$ df $= n_1 + n_2 - 2 =$

Since this is about the difference between two means with equal population variances, the test statistic is:

$$t = \frac{(\bar{x}_1 - \bar{x}_2) - (\mu_1 - \mu_2)}{\sqrt{s_p^2\left(\frac{1}{n_1} + \frac{1}{n_2}\right)}}$$

where $s_p^2 = \frac{(n_1-1)s_1^2+(n_2-1)s_2^2}{n_1+n_2-2}$

Draw rejection region & label critical value(s):

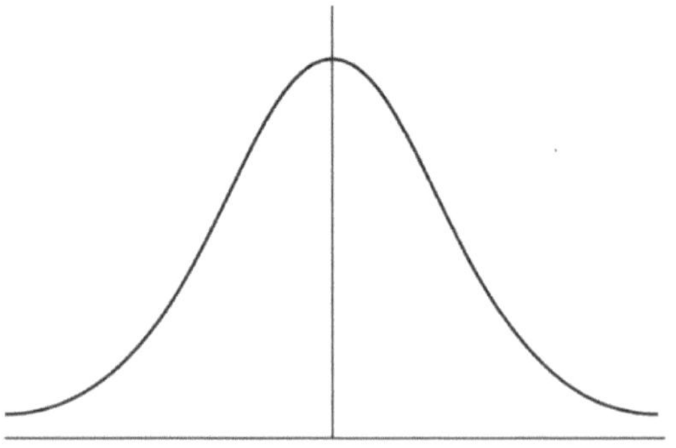

Calculations:

Decision:

Conclusion:

MTH 252 **Names** ____________________

____

GW - Hypothesis Testing - dependent samples ____________________

1. A physical education director claims by taking a special vitamin, a weight lifter can increase his strength. Eight athletes are selected and given a test of strength, using the standard bench press. After 2 weeks of regular training, supplemented with the vitamin, they are tested again. Test the effectiveness of the vitamin regimen at $\alpha = 0.05$. Each value in the table below represents the number of pounds the athlete can bench press. Assume that the variable is approximately normally distributed. Be sure to state how you define your difference, D.

Athlete	**1**	**2**	**3**	**4**	**5**	**6**	**7**	**8**
Before	210	230	182	205	262	253	219	216
After	219	236	179	204	270	250	222	216
Difference								

H_0:

H_1:

(which one is the claim?)

$\overline{D}$ =

s_D =

n =

α = df = $n - 1$ =

Since this is about the mean difference in a dependent sample, the test statistic is:

$$t = \frac{\overline{D} - \mu_D}{s_D/\sqrt{n}}$$

Draw rejection region & label critical value(s):

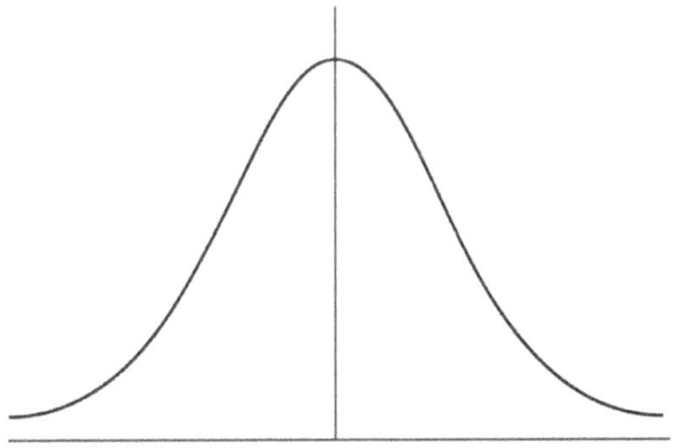

Calculations:

Decision:

Conclusion:

2. An English teacher wishes to see whether a new computer grammar program will reduce the number of grammatical errors her students make when writing a two-page essay. The data are shown here. At 0.025, can it be concluded that the number of errors has been reduced? Be sure to state how you are defining your difference.

Student	A	B	C	D	E	F
Errors before	12	9	0	5	4	3
Errors after	9	6	1	3	2	3
Difference						

H_0:

H_1:

(which one is the claim?)

$\overline{D}$ =

s_D =

n =

α =

df = $n - 1$ =

Since this is about the mean difference in a dependent sample, the test statistic is:

$$t = \frac{\overline{D} - \mu_D}{s_D / \sqrt{n}}$$

Draw rejection region & label critical value(s):

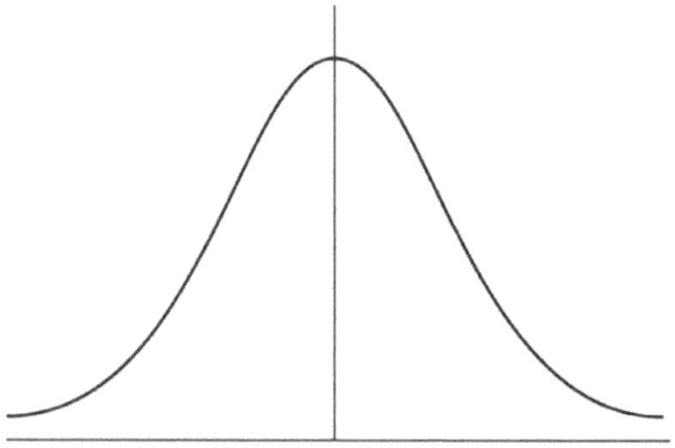

Calculations:

Decision:

Conclusion:

MTH 252	**Names** ______________________
	______________________
GW - Hypothesis Testing - 2 samples – proportions	______________________

1. Bullying is a problem at any age, but especially for students in their teenage years. To see if there is a difference in the rate of bullying between public and private schools, 200 students were randomly selected from each. At the 0.05 level of significance can a difference be concluded?

	Private	**Public**
Sample size	200	200
Number bullied	13	19

H_0:

H_1:

(which one is the claim?)

$p_1 - p_2 =$

$x_1 =$ $x_2 =$

$n_1 =$ $n_2 =$

$\hat{p}_1 =$ $\hat{p}_2 =$

$\alpha =$ $\bar{p} =$

Since this is about the difference between two proportions, the test statistic is:

$$z = \frac{(\hat{p}_1 - \hat{p}_2) - (p_1 - p_2)}{\sqrt{\bar{p}\bar{q}\left(\frac{1}{n_1} + \frac{1}{n_2}\right)}}$$

where $\bar{p} = \frac{x_1+x_2}{n_1+n_2}$ and $\bar{q} = 1 - \bar{p}$

Draw rejection region & label critical value(s):

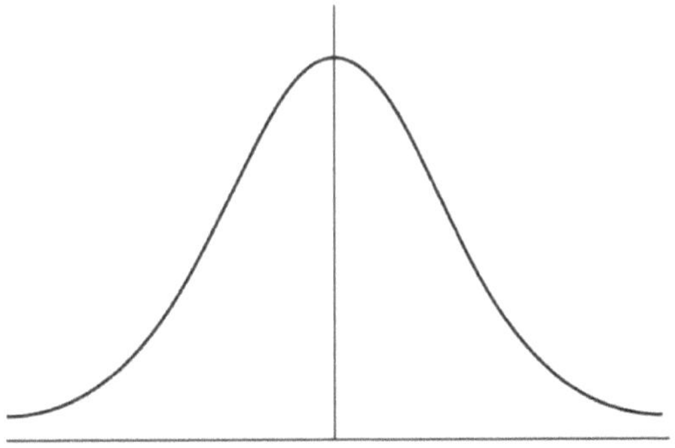

Calculations:

Decision:

Conclusion:

2. In a nursing home study, the researchers found that 12 out of 34 small nursing homes had a resident vaccination rate of less than 80%, while 17 out of 24 large nursing homes had a vaccination rate of less than 80%. At α =0.05, test the claim that there is no difference in the proportions of the small and large nursing homes with a resident vaccination rate of less than 80%.

H_0:	
H_1:	(which one is the claim?)

$p_1 - p_2 =$		Since this is about the difference between two proportions, the test statistic is:
$x_1 =$	$x_2 =$	$$z = \frac{(\hat{p}_1 - \hat{p}_2) - (p_1 - p_2)}{\sqrt{\bar{p}\bar{q}\left(\frac{1}{n_1} + \frac{1}{n_2}\right)}}$$
$n_1 =$	$n_2 =$	
$\hat{p}_1 =$	$\hat{p}_2 =$	where $\bar{p} = \frac{x_1+x_2}{n_1+n_2}$ and $\bar{q} = 1 - \bar{p}$
$\alpha =$	$\bar{p} =$	

Draw rejection region & label critical value(s):

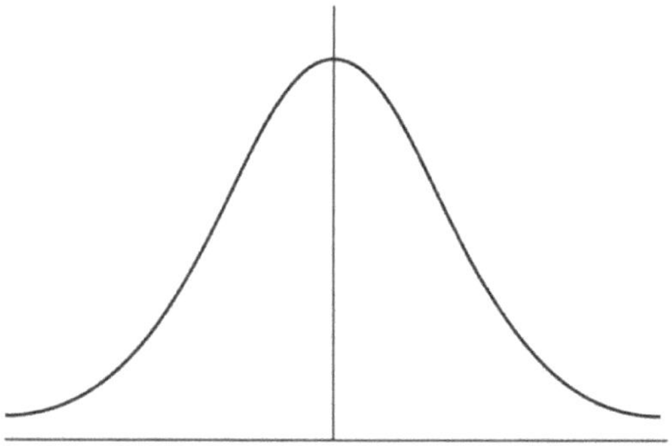

Calculations:

Decision:

Conclusion:

MTH 252	Names ____________________
	____________________
GW - Hypothesis Testing - 2 samples – population variances	____________________

1. A medical researcher wishes to see whether the variance of the heart rates (in beats per minute) of people who vape is different from the variance of heart rates of people who do not vape. Two samples are selected, and the data are as shown. Using α = 0.05, is there enough evidence to support the claim?

Vapers	**Non-vapers**
$n_1 = 26$	$n_2 = 18$
$s_1^2 = 36$	$s_2^2 = 10$

H_0:

H_1: (which one is the claim?)

n_1 =	n_2 =	Since this is about the difference between two variances, the test statistic is:
$s_1^2 =$	$s_2^2 =$	$F = \frac{s_1^2}{s_2^2}$
α =	$dfN =$	
	$dfD =$	where $s_1^2 > s_2^2$

Draw rejection region & label critical value(s):

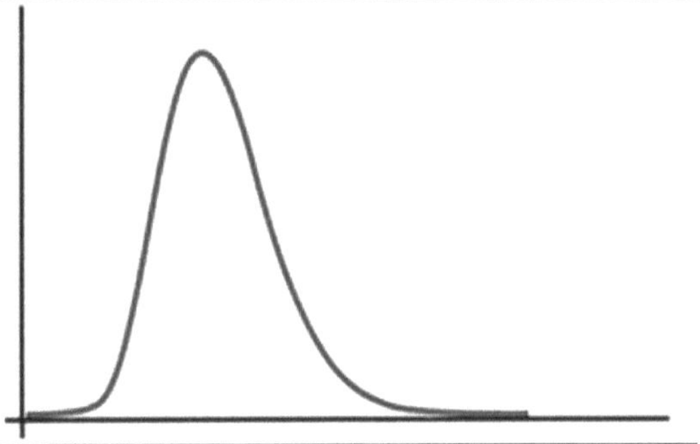

Calculations:

Decision: Conclusion:

2. The standard deviation for the number of weeks 15 fiction books spent on their bestseller list is 6.17 weeks. The standard deviation for 15 nonfiction list is 13.12 weeks. At $\alpha = 0.10$, can we conclude that there is a difference in the variances?

H_0: H_1:	(which one is the claim?)

$n_1 =$	$n_2 =$	Since this is about the difference between two variances, the test statistic is:
$s_1^2 =$	$s_2^2 =$	
$\alpha =$	$dfN =$	$F = \frac{s_1^2}{s_2^2}$
	$dfD =$	where $s_1^2 > s_2^2$

Draw rejection region & label critical value(s):	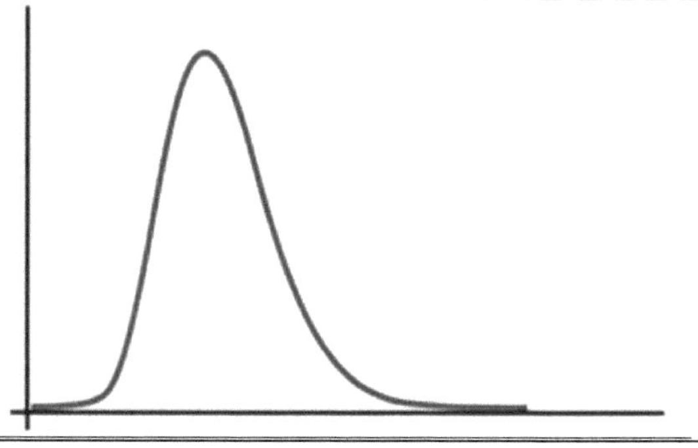

Calculations:

Decision:

Conclusion:

MTH 252	Names ____________________
	____________________
GW - Chi-Square GOF	____________________

1. You would like to determine whether Bristol students have a preference for lunch entrees offered in the cafeteria. You survey 100 students. The results are shown below. At the 0.05 level, do the students indicate a preference for these entrees?

	Chicken Tenders	**Pizza**	**Wraps**	**Burgers**	**Salads**	**Total**
Observed	26	27	12	22	13	100
Expected						

H_0:

H_1:

(which one is the claim?)

df = # categories – 1 =	Since this is about goodness of fit of observed values to hypothesized values, the test statistic is: $$\mathrm{X}^2 = \sum \frac{(O-E)^2}{E}$$

Draw rejection region & label critical value(s):

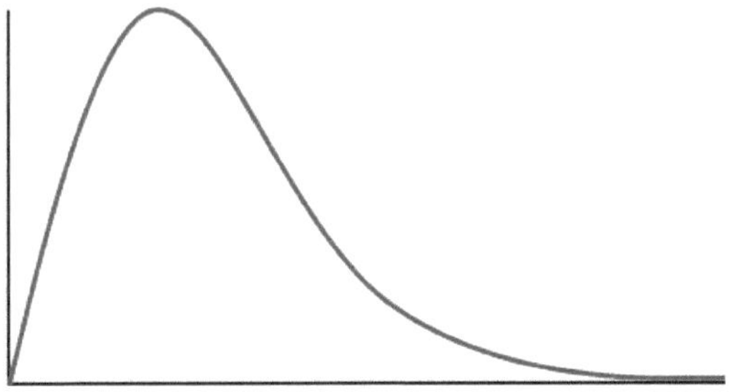

Calculations:

Decision: Conclusion:

2. Human blood is grouped into four types: A, B, AB, and O. The percents of Americans with each type are as follows: O, 43%; A, 40%; B, 12%; and AB, 5%. At a recent blood drive at a large university, the donors were classified as shown below. At the 0.05 level of significance, is there sufficient evidence to conclude that the proportions differ from those stated above?

	O	A	B	AB	Total
Observed	60	65	20	15	160
Expected					160

H_0:

H_1:

(which one is the claim?)

df = # categories – 1 =

Since this is about goodness of fit of observed values to hypothesized values, the test statistic is:

$$X^2 = \sum \frac{(O - E)^2}{E}$$

Draw rejection region & label critical value(s):

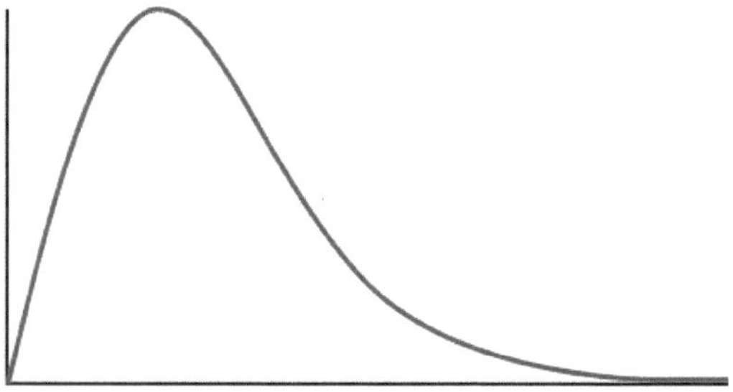

Calculations:

Decision: Conclusion:

MTH 252	Names ____________________
	____________________
GW - Chi-Square Test for Independence	____________________

1. A researcher in the Transfer office wishes to determine whether there is a relationship between the gender of an individual and the academic major chosen. A sample of 100 people is selected, and the following data are obtained.

	Academic Major			
Gender	Business	Nursing	Human Services	Total
Male	20	9	11	40
Female	20	21	19	60
Total	40	30	30	100

At $\alpha = 0.10$, can the researcher conclude that academic major is related to gender?

H_0:

H_1:

(which one is the claim?)

df = (# rows – 1)(#columns – 1) = $Exp = \left(\frac{row\ total}{grand\ total}\right)(column\ total)$	Since this is about goodness of fit of observed values to hypothesized values, the test statistic is: $\mathrm{X}^2 = \sum \frac{(O-E)^2}{E}$

Draw rejection region & label critical value(s):

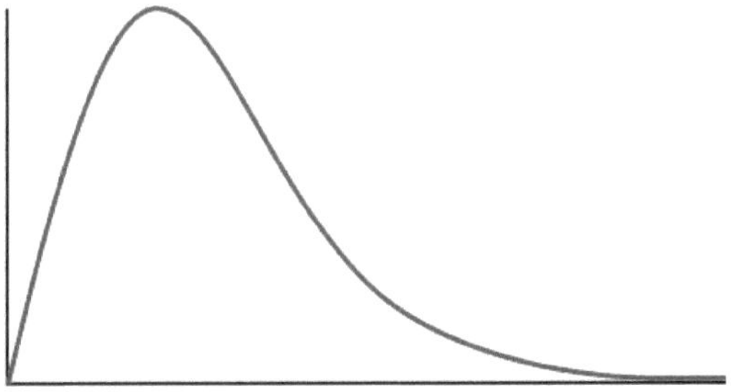

Calculations:

Decision: Conclusion:

2. It is always interesting to see what occupations legislators hold in private life. Can it be concluded that the session of Congress and the occupations of the legislators are dependent? Use $\alpha = 0.05$.

	109th Congress	108th Congress	107th Congress	Total
Agriculture	5	5	6	16
Business	30	25	24	79
Education	12	12	16	40
Law	58	59	53	170
Medicine	4	3	3	10
Total	109	104	102	315

H_0:

H_1:

(which one is the claim?)

df = (# rows – 1)(#columns – 1) =

$$Exp = \left(\frac{row\ total}{grand\ total}\right)(column\ total)$$

Since this is about goodness of fit of observed values to hypothesized values, the test statistic is:

$$X^2 = \sum \frac{(O - E)^2}{E}$$

Draw rejection region & label critical value(s):

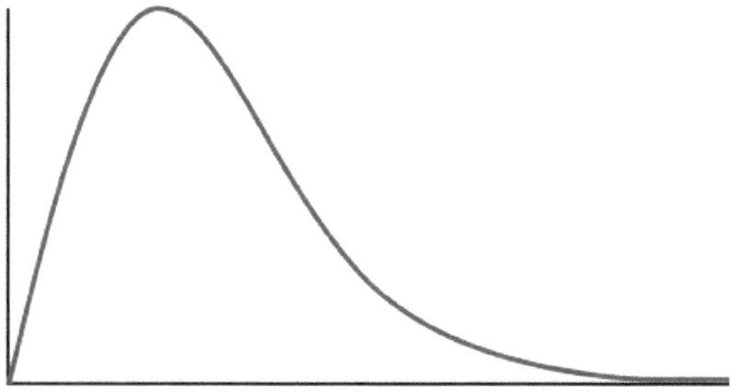

Calculations:

Decision: Conclusion:

MTH 252	**Names** ______________________
GW – Chi-Square Test for	______________________
Homogeneity of Proportions	______________________

1. According to the auto club, 31 million Americans plan to travel over the Thanksgiving holiday. To determine whether to stay open or not, the Benny's restaurant chain surveyed 125 customers at each of four locations to see if they would be traveling over the holiday. The results are shown here. At α= 0.10, test the claim that the proportions of Americans who will travel over the Thanksgiving holiday are equal.

Store Location	A	B	C	D	Total
Will travel	37	52	46	49	184
Will not travel	88	73	79	76	316
Total	125	125	125	125	500

H_0:

H_1:

(which one is the claim?)

df = (# rows – 1)(#columns – 1) =

Since this is about whether the proportions are the same for each location, the test statistic is:

$$X^2 = \sum \frac{(O - E)^2}{E}$$

Draw rejection region & label critical value(s):

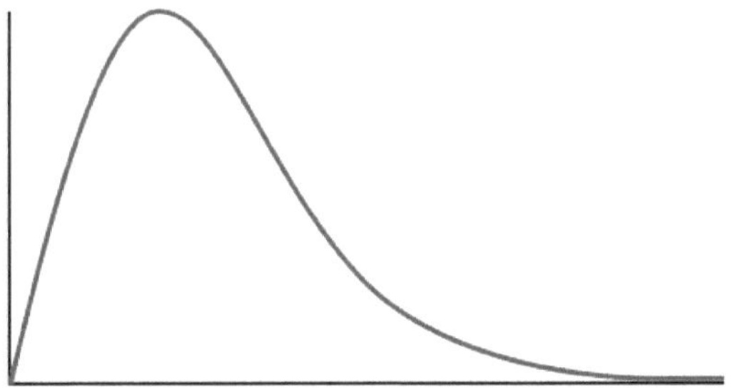

Calculations:

Decision:

Conclusion:

2. An advertising firm has decided to ask 92 customers at each of three local shopping centers if they are willing to participate in a market research survey. According to previous studies, 38% of Americans refuse to take part in such surveys. The results are shown here. At α= 0.01, test the claim that the proportions of those who are willing to participate are equal.

Shopping Center	A	B	C	Total
Will participate	52	45	36	133
Will not participate	40	47	56	143
Total	92	92	92	276

H_0:

H_1:

(which one is the claim?)

df = (# rows – 1)(#columns – 1) =

Since this is about whether the proportions are the same for each center, the test statistic is:

$$X^2 = \sum \frac{(O-E)^2}{E}$$

Draw rejection region & label critical value(s):

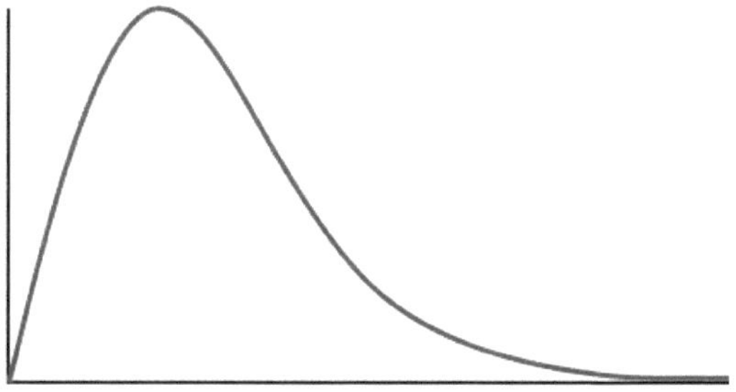

Calculations:

Decision:

Conclusion:

MTH 252	**Names** ____________________
GW - ANOVA	____________________
	____________________

A market researcher wishes to see if there is a significant difference in the number of full-time employees at different chain convenience stores. The data are shown. At $\alpha = 0.05$, can it be concluded that there is a significant difference in the average number of full-time employees at each chain store?

Chain A	**Chain B**	**Chain C**
7	10	1
14	1	12
32	1	1
19	0	9
10	11	1
11	1	11
$\bar{x}_1 = 15.5;\ s_1^2 = 81.9$	$\bar{x}_2 = 4.0;\ s_2^2 = 25.6$	$\bar{x}_3 = 5.8;\ s_3^2 = 29.0$

H_0:

H_1:

(which one is the claim?)

$$s_B^2 = \frac{\sum n_i(\bar{x}_i - \bar{x}_{GM})^2}{k-1}$$

$$s_W^2 = \frac{\sum(n_i - 1)s_i^2}{\sum(n_i - 1)}$$

Since this is testing the equality of multiple means, the test statistic is:

$$F = \frac{s_B^2}{s_W^2}$$

Draw rejection region & label critical value(s):

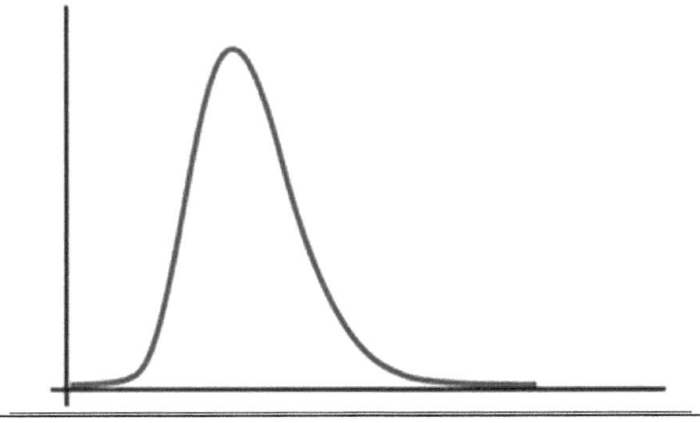

Calculations:

Decision:

Conclusion:

MTH 252	**Names** ______________________
GW – Correlation & Regression	______________________
	______________________

A certain statistics instructor wants to see if there is a relationship between the number of student absences and students' final averages in MTH 252. A random sample of seven students' data is shown below.

Student	**x, # abs.**	**y, final avg.**	$x_i - \bar{x}$	$y_i - \bar{y}$	$(x_i - \bar{x})(y_i - \bar{y})$	$(x_i - \bar{x})^2$
A	6	82				
B	2	86				
C	15	43				
D	9	74				
E	12	58				
F	5	90				
G	8	78				

1. Construct a scatterplot of the data.
2. Fins the equation for the line of best fit. Use $slope = b_1 = \frac{\sum(x_i-\bar{x})(y_i-\bar{y})}{\sum(x_i-\bar{x})^2}$ and $y - imtercept = b_0 = \bar{y} - b_1\bar{x}$. The line of best fit is $\hat{y} = b_1x + b_0$.
3. Find the covariance of this relationship: $COV(x, y) = \frac{\sum(x_i-\bar{x})(y_i-\bar{y})}{n-1}$. Is it positive, negative, or zero?
4. Find the correlation coefficient, r: $r = \frac{COV(x,y)}{s_x \cdot s_y}$. Is it strong/weak, poisitive/negative?
5. Test the statistical significance of r using: $t = r\sqrt{\frac{n-2}{1-r^2}}$. There are $n - 2$ degrees of freedom.
6. Find the coefficient of determination, r^2. What does this tell you about the proportion of variation in y that is explained by the regression (i.e., by x)??

This page intentionally left blank.

The Standard Normal Distribution Table

Cumulative Standard Normal Distribution

z	.00	.01	.02	.03	.04	.05	.06	.07	.08	.09
−3.4	.0003	.0003	.0003	.0003	.0003	.0003	.0003	.0003	.0003	.0002
−3.3	.0005	.0005	.0005	.0004	.0004	.0004	.0004	.0004	.0004	.0003
−3.2	.0007	.0007	.0006	.0006	.0006	.0006	.0006	.0005	.0005	.0005
−3.1	.0010	.0009	.0009	.0009	.0008	.0008	.0008	.0008	.0007	.0007
−3.0	.0013	.0013	.0013	.0012	.0012	.0011	.0011	.0011	.0010	.0010
−2.9	.0019	.0018	.0018	.0017	.0016	.0016	.0015	.0015	.0014	.0014
−2.8	.0026	.0025	.0024	.0023	.0023	.0022	.0021	.0021	.0020	.0019
−2.7	.0035	.0034	.0033	.0032	.0031	.0030	.0029	.0028	.0027	.0026
−2.6	.0047	.0045	.0044	.0043	.0041	.0040	.0039	.0038	.0037	.0036
−2.5	.0062	.0060	.0059	.0057	.0055	.0054	.0052	.0051	.0049	.0048
−2.4	.0082	.0080	.0078	.0075	.0073	.0071	.0069	.0068	.0066	.0064
−2.3	.0107	.0104	.0102	.0099	.0096	.0094	.0091	.0089	.0087	.0084
−2.2	.0139	.0136	.0132	.0129	.0125	.0122	.0119	.0116	.0113	.0110
−2.1	.0179	.0174	.0170	.0166	.0162	.0158	.0154	.0150	.0146	.0143
−2.0	.0228	.0222	.0217	.0212	.0207	.0202	.0197	.0192	.0188	.0183
−1.9	.0287	.0281	.0274	.0268	.0262	.0256	.0250	.0244	.0239	.0233
−1.8	.0359	.0351	.0344	.0336	.0329	.0322	.0314	.0307	.0301	.0294
−1.7	.0446	.0436	.0427	.0418	.0409	.0401	.0392	.0384	.0375	.0367
−1.6	.0548	.0537	.0526	.0516	.0505	.0495	.0485	.0475	.0465	.0455
−1.5	.0668	.0655	.0643	.0630	.0618	.0606	.0594	.0582	.0571	.0559
−1.4	.0808	.0793	.0778	.0764	.0749	.0735	.0721	.0708	.0694	.0681
−1.3	.0968	.0951	.0934	.0918	.0901	.0885	.0869	.0853	.0838	.0823
−1.2	.1151	.1131	.1112	.1093	.1075	.1056	.1038	.1020	.1003	.0985
−1.1	.1357	.1335	.1314	.1292	.1271	.1251	.1230	.1210	.1190	.1170
−1.0	.1587	.1562	.1539	.1515	.1492	.1469	.1446	.1423	.1401	.1379
−0.9	.1841	.1814	.1788	.1762	.1736	.1711	.1685	.1660	.1635	.1611
−0.8	.2119	.2090	.2061	.2033	.2005	.1977	.1949	.1922	.1894	.1867
−0.7	.2420	.2389	.2358	.2327	.2296	.2266	.2236	.2206	.2177	.2148
−0.6	.2743	.2709	.2676	.2643	.2611	.2578	.2546	.2514	.2483	.2451
−0.5	.3085	.3050	.3015	.2981	.2946	.2912	.2877	.2843	.2810	.2776
−0.4	.3446	.3409	.3372	.3336	.3300	.3264	.3228	.3192	.3156	.3121
−0.3	.3821	.3783	.3745	.3707	.3669	.3632	.3594	.3557	.3520	.3483
−0.2	.4207	.4168	.4129	.4090	.4052	.4013	.3974	.3936	.3897	.3859
−0.1	.4602	.4562	.4522	.4483	.4443	.4404	.4364	.4325	.4286	.4247
−0.0	.5000	.4960	.4920	.4880	.4840	.4801	.4761	.4721	.4681	.4641

For z values less than −3.49, use 0.0001.

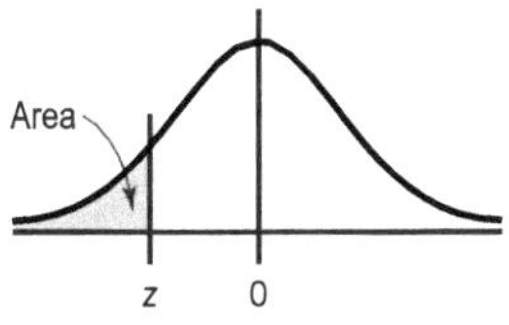

The Standard Normal Distribution Table (continued)

Cumulative Standard Normal Distribution

z	.00	.01	.02	.03	.04	.05	.06	.07	.08	.09
0.0	.5000	.5040	.5080	.5120	.5160	.5199	.5239	.5279	.5319	.5359
0.1	.5398	.5438	.5478	.5517	.5557	.5596	.5636	.5675	.5714	.5753
0.2	.5793	.5832	.5871	.5910	.5948	.5987	.6026	.6064	.6103	.6141
0.3	.6179	.6217	.6255	.6293	.6331	.6368	.6406	.6443	.6480	.6517
0.4	.6554	.6591	.6628	.6664	.6700	.6736	.6772	.6808	.6844	.6879
0.5	.6915	.6950	.6985	.7019	.7054	.7088	.7123	.7157	.7190	.7224
0.6	.7257	.7291	.7324	.7357	.7389	.7422	.7454	.7486	.7517	.7549
0.7	.7580	.7611	.7642	.7673	.7704	.7734	.7764	.7794	.7823	.7852
0.8	.7881	.7910	.7939	.7967	.7995	.8023	.8051	.8078	.8106	.8133
0.9	.8159	.8186	.8212	.8238	.8264	.8289	.8315	.8340	.8365	.8389
1.0	.8413	.8438	.8461	.8485	.8508	.8531	.8554	.8577	.8599	.8621
1.1	.8643	.8665	.8686	.8708	.8729	.8749	.8770	.8790	.8810	.8830
1.2	.8849	.8869	.8888	.8907	.8925	.8944	.8962	.8980	.8997	.9015
1.3	.9032	.9049	.9066	.9082	.9099	.9115	.9131	.9147	.9162	.9177
1.4	.9192	.9207	.9222	.9236	.9251	.9265	.9279	.9292	.9306	.9319
1.5	.9332	.9345	.9357	.9370	.9382	.9394	.9406	.9418	.9429	.9441
1.6	.9452	.9463	.9474	.9484	.9495	.9505	.9515	.9525	.9535	.9545
1.7	.9554	.9564	.9573	.9582	.9591	.9599	.9608	.9616	.9625	.9633
1.8	.9641	.9649	.9656	.9664	.9671	.9678	.9686	.9693	.9699	.9706
1.9	.9713	.9719	.9726	.9732	.9738	.9744	.9750	.9756	.9761	.9767
2.0	.9772	.9778	.9783	.9788	.9793	.9798	.9803	.9808	.9812	.9817
2.1	.9821	.9826	.9830	.9834	.9838	.9842	.9846	.9850	.9854	.9857
2.2	.9861	.9864	.9868	.9871	.9875	.9878	.9881	.9884	.9887	.9890
2.3	.9893	.9896	.9898	.9901	.9904	.9906	.9909	.9911	.9913	.9916
2.4	.9918	.9920	.9922	.9925	.9927	.9929	.9931	.9932	.9934	.9936
2.5	.9938	.9940	.9941	.9943	.9945	.9946	.9948	.9949	.9951	.9952
2.6	.9953	.9955	.9956	.9957	.9959	.9960	.9961	.9962	.9963	.9964
2.7	.9965	.9966	.9967	.9968	.9969	.9970	.9971	.9972	.9973	.9974
2.8	.9974	.9975	.9976	.9977	.9977	.9978	.9979	.9979	.9980	.9981
2.9	.9981	.9982	.9982	.9983	.9984	.9984	.9985	.9985	.9986	.9986
3.0	.9987	.9987	.9987	.9988	.9988	.9989	.9989	.9989	.9990	.9990
3.1	.9990	.9991	.9991	.9991	.9992	.9992	.9992	.9992	.9993	.9993
3.2	.9993	.9993	.9994	.9994	.9994	.9994	.9994	.9995	.9995	.9995
3.3	.9995	.9995	.9995	.9996	.9996	.9996	.9996	.9996	.9996	.9997
3.4	.9997	.9997	.9997	.9997	.9997	.9997	.9997	.9997	.9997	.9998

For z values greater than 3.49, use 0.9999.

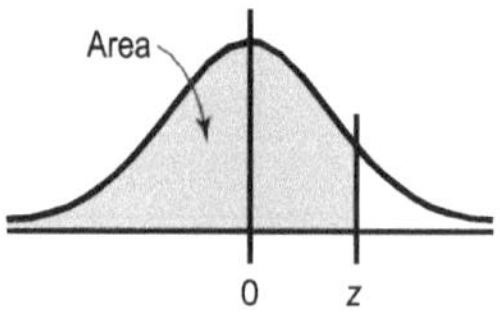

The t Distribution Table

	Confidence intervals	80%	90%	95%	98%	99%
	One tail, α	**0.10**	**0.05**	**0.025**	**0.01**	**0.005**
d.f.	**Two tails, α**	**0.20**	**0.10**	**0.05**	**0.02**	**0.01**
1		3.078	6.314	12.706	31.821	63.657
2		1.886	2.920	4.303	6.965	9.925
3		1.638	2.353	3.182	4.541	5.841
4		1.533	2.132	2.776	3.747	4.604
5		1.476	2.015	2.571	3.365	4.032
6		1.440	1.943	2.447	3.143	3.707
7		1.415	1.895	2.365	2.998	3.499
8		1.397	1.860	2.306	2.896	3.355
9		1.383	1.833	2.262	2.821	3.250
10		1.372	1.812	2.228	2.764	3.169
11		1.363	1.796	2.201	2.718	3.106
12		1.356	1.782	2.179	2.681	3.055
13		1.350	1.771	2.160	2.650	3.012
14		1.345	1.761	2.145	2.624	2.977
15		1.341	1.753	2.131	2.602	2.947
16		1.337	1.746	2.120	2.583	2.921
17		1.333	1.740	2.110	2.567	2.898
18		1.330	1.734	2.101	2.552	2.878
19		1.328	1.729	2.093	2.539	2.861
20		1.325	1.725	2.086	2.528	2.845
21		1.323	1.721	2.080	2.518	2.831
22		1.321	1.717	2.074	2.508	2.819
23		1.319	1.714	2.069	2.500	2.807
24		1.318	1.711	2.064	2.492	2.797
25		1.316	1.708	2.060	2.485	2.787
26		1.315	1.706	2.056	2.479	2.779
27		1.314	1.703	2.052	2.473	2.771
28		1.313	1.701	2.048	2.467	2.763
29		1.311	1.699	2.045	2.462	2.756
30		1.310	1.697	2.042	2.457	2.750
32		1.309	1.694	2.037	2.449	2.738
34		1.307	1.691	2.032	2.441	2.728
36		1.306	1.688	2.028	2.434	2.719
38		1.304	1.686	2.024	2.429	2.712
40		1.303	1.684	2.021	2.423	2.704
45		1.301	1.679	2.014	2.412	2.690
50		1.299	1.676	2.009	2.403	2.678
55		1.297	1.673	2.004	2.396	2.668
60		1.296	1.671	2.000	2.390	2.660
65		1.295	1.669	1.997	2.385	2.654
70		1.294	1.667	1.994	2.381	2.648
75		1.293	1.665	1.992	2.377	2.643
80		1.292	1.664	1.990	2.374	2.639
90		1.291	1.662	1.987	2.368	2.632
100		1.290	1.660	1.984	2.364	2.626
500		1.283	1.648	1.965	2.334	2.586
1000		1.282	1.646	1.962	2.330	2.581
(z) ∞		1.282[a]	1.645[b]	1.960	2.326[c]	2.576[d]

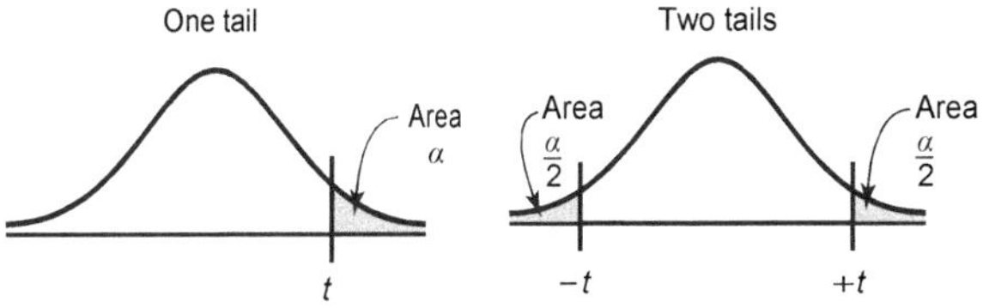

The Chi-Square Distribution Table

Degrees of freedom	α 0.995	0.99	0.975	0.95	0.90	0.10	0.05	0.025	0.01	0.005
1	—	—	0.001	0.004	0.016	2.706	3.841	5.024	6.635	7.879
2	0.010	0.020	0.051	0.103	0.211	4.605	5.991	7.378	9.210	10.597
3	0.072	0.115	0.216	0.352	0.584	6.251	7.815	9.348	11.345	12.838
4	0.207	0.297	0.484	0.711	1.064	7.779	9.488	11.143	13.277	14.860
5	0.412	0.554	0.831	1.145	1.610	9.236	11.071	12.833	15.086	16.750
6	0.676	0.872	1.237	1.635	2.204	10.645	12.592	14.449	16.812	18.548
7	0.989	1.239	1.690	2.167	2.833	12.017	14.067	16.013	18.475	20.278
8	1.344	1.646	2.180	2.733	3.490	13.362	15.507	17.535	20.090	21.955
9	1.735	2.088	2.700	3.325	4.168	14.684	16.919	19.023	21.666	23.589
10	2.156	2.558	3.247	3.940	4.865	15.987	18.307	20.483	23.209	25.188
11	2.603	3.053	3.816	4.575	5.578	17.275	19.675	21.920	24.725	26.757
12	3.074	3.571	4.404	5.226	6.304	18.549	21.026	23.337	26.217	28.299
13	3.565	4.107	5.009	5.892	7.042	19.812	22.362	24.736	27.688	29.819
14	4.075	4.660	5.629	6.571	7.790	21.064	23.685	26.119	29.141	31.319
15	4.601	5.229	6.262	7.261	8.547	22.307	24.996	27.488	30.578	32.801
16	5.142	5.812	6.908	7.962	9.312	23.542	26.296	28.845	32.000	34.267
17	5.697	6.408	7.564	8.672	10.085	24.769	27.587	30.191	33.409	35.718
18	6.265	7.015	8.231	9.390	10.865	25.989	28.869	31.526	34.805	37.156
19	6.844	7.633	8.907	10.117	11.651	27.204	30.144	32.852	36.191	38.582
20	7.434	8.260	9.591	10.851	12.443	28.412	31.410	34.170	37.566	39.997
21	8.034	8.897	10.283	11.591	13.240	29.615	32.671	35.479	38.932	41.401
22	8.643	9.542	10.982	12.338	14.042	30.813	33.924	36.781	40.289	42.796
23	9.262	10.196	11.689	13.091	14.848	32.007	35.172	38.076	41.638	44.181
24	9.886	10.856	12.401	13.848	15.659	33.196	36.415	39.364	42.980	45.559
25	10.520	11.524	13.120	14.611	16.473	34.382	37.652	40.646	44.314	46.928
26	11.160	12.198	13.844	15.379	17.292	35.563	38.885	41.923	45.642	48.290
27	11.808	12.879	14.573	16.151	18.114	36.741	40.113	43.194	46.963	49.645
28	12.461	13.565	15.308	16.928	18.939	37.916	41.337	44.461	48.278	50.993
29	13.121	14.257	16.047	17.708	19.768	39.087	42.557	45.722	49.588	52.336
30	13.787	14.954	16.791	18.493	20.599	40.256	43.773	46.979	50.892	53.672
40	20.707	22.164	24.433	26.509	29.051	51.805	55.758	59.342	63.691	66.766
50	27.991	29.707	32.357	34.764	37.689	63.167	67.505	71.420	76.154	79.490
60	35.534	37.485	40.482	43.188	46.459	74.397	79.082	83.298	88.379	91.952
70	43.275	45.442	48.758	51.739	55.329	85.527	90.531	95.023	100.425	104.215
80	51.172	53.540	57.153	60.391	64.278	96.578	101.879	106.629	112.329	116.321
90	59.196	61.754	65.647	69.126	73.291	107.565	113.145	118.136	124.116	128.299
100	67.328	70.065	74.222	77.929	82.358	118.498	124.342	129.561	135.807	140.169

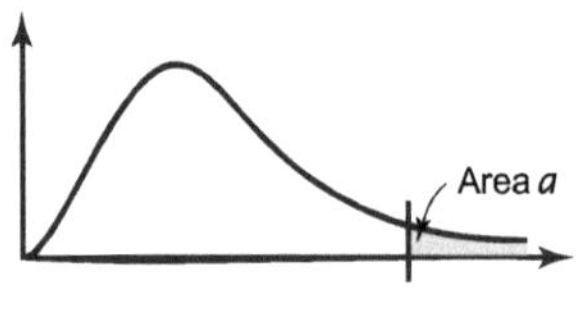

F Distribution table [alpha=0.05]

/	$df_1=1$	2	3	4	5	6	7	8	9	10	12	15	20	24	30	40	60	120	inf
$df_2=1$	161.4	199.5	215.7	224.6	230.2	234.0	236.8	238.9	240.5	241.9	243.9	245.9	248.0	249.1	250.1	251.1	252.2	253.3	254.3
2	18.51	19.00	19.16	19.25	19.30	19.33	19.35	19.37	19.38	19.40	19.41	19.43	19.45	19.45	19.46	19.47	19.48	19.49	19.50
3	10.13	9.552	9.277	9.117	9.014	8.941	8.887	8.845	8.812	8.786	8.745	8.703	8.660	8.639	8.617	8.594	8.572	8.549	8.526
4	7.709	6.944	6.591	6.388	6.256	6.163	6.094	6.041	5.999	5.964	5.912	5.858	5.803	5.774	5.746	5.717	5.688	5.658	5.628
5	6.608	5.786	5.409	5.192	5.050	4.950	4.876	4.818	4.772	4.735	4.678	4.619	4.558	4.527	4.496	4.464	4.431	4.398	4.365
6	5.987	5.143	4.757	4.534	4.387	4.284	4.207	4.147	4.099	4.060	4.000	3.938	3.874	3.841	3.808	3.774	3.740	3.705	3.669
7	5.591	4.737	4.347	4.120	3.972	3.866	3.787	3.726	3.677	3.637	3.575	3.511	3.445	3.410	3.376	3.340	3.304	3.267	3.230
8	5.318	4.459	4.066	3.838	3.688	3.581	3.500	3.438	3.388	3.347	3.284	3.218	3.150	3.115	3.079	3.043	3.005	2.967	2.928
9	5.117	4.256	3.863	3.633	3.482	3.374	3.293	3.230	3.179	3.137	3.073	3.006	2.936	2.900	2.864	2.826	2.787	2.748	2.707
10	4.965	4.103	3.708	3.478	3.326	3.217	3.135	3.072	3.020	2.978	2.913	2.845	2.774	2.737	2.700	2.661	2.621	2.580	2.538
11	4.844	3.982	3.587	3.357	3.204	3.095	3.012	2.948	2.896	2.854	2.788	2.719	2.646	2.609	2.570	2.531	2.490	2.448	2.404

/	df_1=1	2	3	4	5	6	7	8	9	10	12	15	20	24	30	40	60	120	inf
12	4.747	3.885	3.490	3.259	3.106	2.996	2.913	2.849	2.796	2.753	2.687	2.617	2.544	2.505	2.466	2.426	2.384	2.341	2.296
13	4.667	3.806	3.411	3.179	3.025	2.915	2.832	2.767	2.714	2.671	2.604	2.533	2.459	2.420	2.380	2.339	2.297	2.252	2.206
14	4.600	3.739	3.344	3.112	2.958	2.848	2.764	2.699	2.646	2.602	2.534	2.463	2.388	2.349	2.308	2.266	2.223	2.178	2.131
15	4.543	3.682	3.287	3.056	2.901	2.790	2.707	2.641	2.588	2.544	2.475	2.403	2.328	2.288	2.247	2.204	2.160	2.114	2.066
16	4.494	3.634	3.239	3.007	2.852	2.741	2.657	2.591	2.538	2.494	2.425	2.352	2.276	2.235	2.194	2.151	2.106	2.059	2.010
17	4.451	3.592	3.197	2.965	2.810	2.699	2.614	2.548	2.494	2.450	2.381	2.308	2.230	2.190	2.148	2.104	2.058	2.011	1.960
18	4.414	3.555	3.160	2.928	2.773	2.661	2.577	2.510	2.456	2.412	2.342	2.269	2.191	2.150	2.107	2.063	2.017	1.968	1.917
19	4.381	3.522	3.127	2.895	2.740	2.628	2.544	2.477	2.423	2.378	2.308	2.234	2.156	2.114	2.071	2.026	1.980	1.930	1.878
20	4.351	3.493	3.098	2.866	2.711	2.599	2.514	2.447	2.393	2.348	2.278	2.203	2.124	2.082	2.039	1.994	1.946	1.896	1.843
21	4.325	3.467	3.072	2.840	2.685	2.573	2.488	2.420	2.366	2.321	2.250	2.176	2.096	2.054	2.010	1.965	1.916	1.866	1.812
22	4.301	3.443	3.049	2.817	2.661	2.549	2.464	2.397	2.342	2.297	2.226	2.151	2.071	2.028	1.984	1.938	1.889	1.838	1.783
23	4.279	3.422	3.028	2.796	2.640	2.528	2.442	2.375	2.320	2.275	2.204	2.128	2.048	2.005	1.961	1.914	1.865	1.813	1.757

/	df_1=1	2	3	4	5	6	7	8	9	10	12	15	20	24	30	40	60	120	inf
24	4.260	3.403	3.009	2.776	2.621	2.508	2.423	2.355	2.300	2.255	2.183	2.108	2.027	1.984	1.939	1.892	1.842	1.790	1.733
25	4.242	3.385	2.991	2.759	2.603	2.490	2.405	2.337	2.282	2.236	2.165	2.089	2.007	1.964	1.919	1.872	1.822	1.768	1.711
26	4.225	3.369	2.975	2.743	2.587	2.474	2.388	2.321	2.265	2.220	2.148	2.072	1.990	1.946	1.901	1.853	1.803	1.749	1.691
27	4.210	3.354	2.960	2.728	2.572	2.459	2.373	2.305	2.250	2.204	2.132	2.056	1.974	1.930	1.884	1.836	1.785	1.731	1.672
28	4.196	3.340	2.947	2.714	2.558	2.445	2.359	2.291	2.236	2.190	2.118	2.041	1.959	1.915	1.869	1.820	1.769	1.714	1.654
29	4.183	3.328	2.934	2.701	2.545	2.432	2.346	2.278	2.223	2.177	2.104	2.027	1.945	1.901	1.854	1.806	1.754	1.698	1.638
30	4.171	3.316	2.922	2.690	2.534	2.421	2.334	2.266	2.211	2.165	2.092	2.015	1.932	1.887	1.841	1.792	1.740	1.683	1.622
40	4.085	3.232	2.839	2.606	2.449	2.336	2.249	2.180	2.124	2.077	2.003	1.924	1.839	1.793	1.744	1.693	1.637	1.577	1.509
60	4.001	3.150	2.758	2.525	2.368	2.254	2.167	2.097	2.040	1.993	1.917	1.836	1.748	1.700	1.649	1.594	1.534	1.467	1.389
120	3.920	3.072	2.680	2.447	2.290	2.175	2.087	2.016	1.959	1.910	1.834	1.751	1.659	1.608	1.554	1.495	1.429	1.352	1.254
inf	3.842	2.996	2.605	2.372	2.214	2.099	2.010	1.938	1.880	1.831	1.752	1.666	1.571	1.517	1.459	1.394	1.318	1.221	1.000

F Distribution table [alpha=0.025]

/	df_1=1	2	3	4	5	6	7	8	9	10	12	15	20	24	30	40	60	120	8
df_2= 1	647.7890	799.5000	864.1630	899.5833	921.8479	937.1111	948.2169	956.6562	963.2846	968.6274	976.7079	984.8668	993.1028	997.2492	1001.414	1005.598	1009.800	1014.020	1018.258
2	38.5063	39.0000	39.1655	39.2484	39.2982	39.3315	39.3552	39.3730	39.3869	39.3980	39.4146	39.4313	39.4479	39.4562	39.465	39.473	39.481	39.490	39.498
3	17.4434	16.0441	15.4392	15.1010	14.8848	14.7347	14.6244	14.5399	14.4731	14.4189	14.3366	14.2527	14.1674	14.1241	14.081	14.037	13.992	13.947	13.902
4	12.2179	10.6491	9.9792	9.6045	9.3645	9.1973	9.0741	8.9796	8.9047	8.8439	8.7512	8.6565	8.5599	8.5109	8.461	8.411	8.360	8.309	8.257
5	10.0070	8.4336	7.7636	7.3879	7.1464	6.9777	6.8531	6.7572	6.6811	6.6192	6.5245	6.4277	6.3286	6.2780	6.227	6.175	6.123	6.069	6.015
6	8.8131	7.2599	6.5988	6.2272	5.9876	5.8198	5.6955	5.5996	5.5234	5.4613	5.3662	5.2687	5.1684	5.1172	5.065	5.012	4.959	4.904	4.849
7	8.0727	6.5415	5.8898	5.5226	5.2852	5.1186	4.9949	4.8993	4.8232	4.7611	4.6658	4.5678	4.4667	4.4150	4.362	4.309	4.254	4.199	4.142
8	7.5709	6.0595	5.4160	5.0526	4.8173	4.6517	4.5286	4.4333	4.3572	4.2951	4.1997	4.1012	3.9995	3.9472	3.894	3.840	3.784	3.728	3.670
9	7.2093	5.7147	5.0781	4.7181	4.4844	4.3197	4.1970	4.1020	4.0260	3.9639	3.8682	3.7694	3.6669	3.6142	3.560	3.505	3.449	3.392	3.333

/	df_1=1	2	3	4	5	6	7	8	9	10	12	15	20	24	30	40	60	120	8
10	6.9367	5.4564	4.8256	4.4683	4.2361	4.0721	3.9498	3.8549	3.7790	3.7168	3.6209	3.5217	3.4185	3.3654	3.311	3.255	3.198	3.140	3.080
11	6.7241	5.2559	4.6300	4.2751	4.0440	3.8807	3.7586	3.6638	3.5879	3.5257	3.4296	3.3299	3.2261	3.1725	3.118	3.061	3.004	2.944	2.883
12	6.5538	5.0959	4.4742	4.1212	3.8911	3.7283	3.6065	3.5118	3.4358	3.3736	3.2773	3.1772	3.0728	3.0187	2.963	2.906	2.848	2.787	2.725
13	6.4143	4.9653	4.3472	3.9959	3.7667	3.6043	3.4827	3.3880	3.3120	3.2497	3.1532	3.0527	2.9477	2.8932	2.837	2.780	2.720	2.659	2.595
14	6.2979	4.8567	4.2417	3.8919	3.6634	3.5014	3.3799	3.2853	3.2093	3.1469	3.0502	2.9493	2.8437	2.7888	2.732	2.674	2.614	2.552	2.487
15	6.1995	4.7650	4.1528	3.8043	3.5764	3.4147	3.2934	3.1987	3.1227	3.0602	2.9633	2.8621	2.7559	2.7006	2.644	2.585	2.524	2.461	2.395
16	6.1151	4.6867	4.0768	3.7294	3.5021	3.3406	3.2194	3.1248	3.0488	2.9862	2.8890	2.7875	2.6808	2.6252	2.568	2.509	2.447	2.383	2.316
17	6.0420	4.6189	4.0112	3.6648	3.4379	3.2767	3.1556	3.0610	2.9849	2.9222	2.8249	2.7230	2.6158	2.5598	2.502	2.442	2.380	2.315	2.247
18	5.9781	4.5597	3.9539	3.6083	3.3820	3.2209	3.0999	3.0053	2.9291	2.8664	2.7689	2.6667	2.5590	2.5027	2.445	2.384	2.321	2.256	2.187
19	5.9216	4.5075	3.9034	3.5587	3.3327	3.1718	3.0509	2.9563	2.8801	2.8172	2.7196	2.6171	2.5089	2.4523	2.394	2.333	2.270	2.203	2.133

/	df_1=1	2	3	4	5	6	7	8	9	10	12	15	20	24	30	40	60	120	8
20	5.8715	4.4613	3.8587	3.5147	3.2891	3.1283	3.0074	2.9128	2.8365	2.7737	2.6758	2.5731	2.4645	2.4076	2.349	2.287	2.223	2.156	2.085
21	5.8266	4.4199	3.8188	3.4754	3.2501	3.0895	2.9686	2.8740	2.7977	2.7348	2.6368	2.5338	2.4247	2.3675	2.308	2.246	2.182	2.114	2.042
22	5.7863	4.3828	3.7829	3.4401	3.2151	3.0546	2.9338	2.8392	2.7628	2.6998	2.6017	2.4984	2.3890	2.3315	2.272	2.210	2.145	2.076	2.003
23	5.7498	4.3492	3.7505	3.4083	3.1835	3.0232	2.9023	2.8077	2.7313	2.6682	2.5699	2.4665	2.3567	2.2989	2.239	2.176	2.111	2.041	1.968
24	5.7166	4.3187	3.7211	3.3794	3.1548	2.9946	2.8738	2.7791	2.7027	2.6396	2.5411	2.4374	2.3273	2.2693	2.209	2.146	2.080	2.010	1.935
25	5.6864	4.2909	3.6943	3.3530	3.1287	2.9685	2.8478	2.7531	2.6766	2.6135	2.5149	2.4110	2.3005	2.2422	2.182	2.118	2.052	1.981	1.906
26	5.6586	4.2655	3.6697	3.3289	3.1048	2.9447	2.8240	2.7293	2.6528	2.5896	2.4908	2.3867	2.2759	2.2174	2.157	2.093	2.026	1.954	1.878
27	5.6331	4.2421	3.6472	3.3067	3.0828	2.9228	2.8021	2.7074	2.6309	2.5676	2.4688	2.3644	2.2533	2.1946	2.133	2.069	2.002	1.930	1.853
28	5.6096	4.2205	3.6264	3.2863	3.0626	2.9027	2.7820	2.6872	2.6106	2.5473	2.4484	2.3438	2.2324	2.1735	2.112	2.048	1.980	1.907	1.829
29	5.5878	4.2006	3.6072	3.2674	3.0438	2.8840	2.7633	2.6686	2.5919	2.5286	2.4295	2.3248	2.2131	2.1540	2.092	2.028	1.959	1.886	1.807

/	df_1=1	2	3	4	5	6	7	8	9	10	12	15	20	24	30	40	60	120	8
30	5.5675	4.1821	3.5894	3.2499	3.0265	2.8667	2.7460	2.6513	2.5746	2.5112	2.4120	2.3072	2.1952	2.1359	2.074	2.009	1.940	1.866	1.787
40	5.4239	4.0510	3.4633	3.1261	2.9037	2.7444	2.6238	2.5289	2.4519	2.3882	2.2882	2.1819	2.0677	2.0069	1.943	1.875	1.803	1.724	1.637
60	5.2856	3.9253	3.3425	3.0077	2.7863	2.6274	2.5068	2.4117	2.3344	2.2702	2.1692	2.0613	1.9445	1.8817	1.815	1.744	1.667	1.581	1.482
120	5.1523	3.8046	3.2269	2.8943	2.6740	2.5154	2.3948	2.2994	2.2217	2.1570	2.0548	1.9450	1.8249	1.7597	1.690	1.614	1.530	1.433	1.310
8	5.0239	3.6889	3.1161	2.7858	2.5665	2.4082	2.2875	2.1918	2.1136	2.0483	1.9447	1.8326	1.7085	1.6402	1.566	1.484	1.388	1.268	1.000

F Distribution table [alpha=0.01]

/	df_1=1	2	3	4	5	6	7	8	9	10	12	15	20	24	30	40	60	120	8
df_2= 1	4052.181	4999.500	5403.352	5624.583	5763.650	5858.986	5928.356	5981.070	6022.473	6055.847	6106.321	6157.285	6208.730	6234.631	6260.649	6286.782	6313.030	6339.391	6365.864
2	98.503	99.000	99.166	99.249	99.299	99.333	99.356	99.374	99.388	99.399	99.416	99.433	99.449	99.458	99.466	99.474	99.482	99.491	99.499
3	34.116	30.817	29.457	28.710	28.237	27.911	27.672	27.489	27.345	27.229	27.052	26.872	26.690	26.598	26.505	26.411	26.316	26.221	26.125
4	21.198	18.000	16.694	15.977	15.522	15.207	14.976	14.799	14.659	14.546	14.374	14.198	14.020	13.929	13.838	13.745	13.652	13.558	13.463
5	16.258	13.274	12.060	11.392	10.967	10.672	10.456	10.289	10.158	10.051	9.888	9.722	9.553	9.466	9.379	9.291	9.202	9.112	9.020
6	13.745	10.925	9.780	9.148	8.746	8.466	8.260	8.102	7.976	7.874	7.718	7.559	7.396	7.313	7.229	7.143	7.057	6.969	6.880
7	12.246	9.547	8.451	7.847	7.460	7.191	6.993	6.840	6.719	6.620	6.469	6.314	6.155	6.074	5.992	5.908	5.824	5.737	5.650
8	11.259	8.649	7.591	7.006	6.632	6.371	6.178	6.029	5.911	5.814	5.667	5.515	5.359	5.279	5.198	5.116	5.032	4.946	4.859
9	10.561	8.022	6.992	6.422	6.057	5.802	5.613	5.467	5.351	5.257	5.111	4.962	4.808	4.729	4.649	4.567	4.483	4.398	4.311
10	10.044	7.559	6.552	5.994	5.636	5.386	5.200	5.057	4.942	4.849	4.706	4.558	4.405	4.327	4.247	4.165	4.082	3.996	3.909

/	df_1=1	2	3	4	5	6	7	8	9	10	12	15	20	24	30	40	60	120	∞
11	9.646	7.206	6.217	5.668	5.316	5.069	4.886	4.744	4.632	4.539	4.397	4.251	4.099	4.021	3.941	3.860	3.776	3.690	3.602
12	9.330	6.927	5.953	5.412	5.064	4.821	4.640	4.499	4.388	4.296	4.155	4.010	3.858	3.780	3.701	3.619	3.535	3.449	3.361
13	9.074	6.701	5.739	5.205	4.862	4.620	4.441	4.302	4.191	4.100	3.960	3.815	3.665	3.587	3.507	3.425	3.341	3.255	3.165
14	8.862	6.515	5.564	5.035	4.695	4.456	4.278	4.140	4.030	3.939	3.800	3.656	3.505	3.427	3.348	3.266	3.181	3.094	3.004
15	8.683	6.359	5.417	4.893	4.556	4.318	4.142	4.004	3.895	3.805	3.666	3.522	3.372	3.294	3.214	3.132	3.047	2.959	2.868
16	8.531	6.226	5.292	4.773	4.437	4.202	4.026	3.890	3.780	3.691	3.553	3.409	3.259	3.181	3.101	3.018	2.933	2.845	2.753
17	8.400	6.112	5.185	4.669	4.336	4.102	3.927	3.791	3.682	3.593	3.455	3.312	3.162	3.084	3.003	2.920	2.835	2.746	2.653
18	8.285	6.013	5.092	4.579	4.248	4.015	3.841	3.705	3.597	3.508	3.371	3.227	3.077	2.999	2.919	2.835	2.749	2.660	2.566
19	8.185	5.926	5.010	4.500	4.171	3.939	3.765	3.631	3.523	3.434	3.297	3.153	3.003	2.925	2.844	2.761	2.674	2.584	2.489
20	8.096	5.849	4.938	4.431	4.103	3.871	3.699	3.564	3.457	3.368	3.231	3.088	2.938	2.859	2.778	2.695	2.608	2.517	2.421

/	df_1=1	2	3	4	5	6	7	8	9	10	12	15	20	24	30	40	60	120	8
21	8.017	5.780	4.874	4.369	4.042	3.812	3.640	3.506	3.398	3.310	3.173	3.030	2.880	2.801	2.720	2.636	2.548	2.457	2.360
22	7.945	5.719	4.817	4.313	3.988	3.758	3.587	3.453	3.346	3.258	3.121	2.978	2.827	2.749	2.667	2.583	2.495	2.403	2.305
23	7.881	5.664	4.765	4.264	3.939	3.710	3.539	3.406	3.299	3.211	3.074	2.931	2.781	2.702	2.620	2.535	2.447	2.354	2.256
24	7.823	5.614	4.718	4.218	3.895	3.667	3.496	3.363	3.256	3.168	3.032	2.889	2.738	2.659	2.577	2.492	2.403	2.310	2.211
25	7.770	5.568	4.675	4.177	3.855	3.627	3.457	3.324	3.217	3.129	2.993	2.850	2.699	2.620	2.538	2.453	2.364	2.270	2.169
26	7.721	5.526	4.637	4.140	3.818	3.591	3.421	3.288	3.182	3.094	2.958	2.815	2.664	2.585	2.503	2.417	2.327	2.233	2.131
27	7.677	5.488	4.601	4.106	3.785	3.558	3.388	3.256	3.149	3.062	2.926	2.783	2.632	2.552	2.470	2.384	2.294	2.198	2.097
28	7.636	5.453	4.568	4.074	3.754	3.528	3.358	3.226	3.120	3.032	2.896	2.753	2.602	2.522	2.440	2.354	2.263	2.167	2.064
29	7.598	5.420	4.538	4.045	3.725	3.499	3.330	3.198	3.092	3.005	2.868	2.726	2.574	2.495	2.412	2.325	2.234	2.138	2.034
30	7.562	5.390	4.510	4.018	3.699	3.473	3.304	3.173	3.067	2.979	2.843	2.700	2.549	2.469	2.386	2.299	2.208	2.111	2.006

Studentized Range Table (α = 0.05)

Critical values of the studentized range for α = 0.05.

					D_1				
D_2	2	3	4	5	6	7	8	9	10
1	17.97	26.98	38.32	37.08	40.41	43.12	45.40	47.36	49.07
2	6.09	8.33	9.80	10.88	11.74	12.44	13.03	13.54	13.99
3	4.50	5.91	6.82	7.50	8.04	8.48	8.85	9.18	9.46
4	3.93	5.04	5.76	6.29	6.71	7.05	7.35	7.60	7.83
5	3.64	4.60	5.22	5.67	6.03	6.33	6.58	6.80	6.99
6	3.46	4.34	4.90	5.30	5.63	5.90	6.12	6.32	6.49
7	3.34	4.16	4.68	5.06	5.36	5.61	5.82	6.00	6.16
8	3.26	4.04	4.53	4.89	5.17	5.40	5.60	5.77	5.92
9	3.20	3.95	4.41	4.76	5.02	5.24	5.43	5.59	5.74
10	3.15	3.88	4.33	4.65	4.91	5.12	5.30	5.46	5.60
11	3.11	3.82	4.26	4.57	4.82	5.03	5.20	5.35	5.49
12	3.08	3.77	4.20	4.51	4.75	4.95	5.12	5.27	5.39
13	3.06	3.73	4.15	4.45	4.69	4.88	5.05	5.19	5.32
14	3.03	3.70	4.11	4.41	4.64	4.83	4.99	5.13	5.25
15	3.01	3.67	4.08	4.37	4.59	4.78	4.94	5.08	5.20
16	3.00	3.65	4.05	4.33	4.56	4.74	4.90	5.03	5.15
17	2.98	3.63	4.02	4.30	4.52	4.70	4.86	4.99	5.11

/	df_1=1	2	3	4	5	6	7	8	9	10	12	15	20	24	30	40	60	120	8
40	7.314	5.179	4.313	3.828	3.514	3.291	3.124	2.993	2.888	2.801	2.665	2.522	2.369	2.288	2.203	2.114	2.019	1.917	1.805
60	7.077	4.977	4.126	3.649	3.339	3.119	2.953	2.823	2.718	2.632	2.496	2.352	2.198	2.115	2.028	1.936	1.836	1.726	1.601
120	6.851	4.787	3.949	3.480	3.174	2.956	2.792	2.663	2.559	2.472	2.336	2.192	2.035	1.950	1.860	1.763	1.656	1.533	1.381
8	6.635	4.605	3.782	3.319	3.017	2.802	2.639	2.511	2.407	2.321	2.185	2.039	1.878	1.791	1.696	1.592	1.473	1.325	1.000